FRESH WATER ECO-SYSTEM

A Case Study of River Henwal

FRESH WATER ECO-SYSTEM

A Case Study of River Henwal

D.R. Khanna
Gagan Matta

BIOTECH
2011
BIOTECH BOOKS

Prof. D.R. Khanna
Prof. Gagan Matta

ISBN 978-81-7622-224-2

Published by
BIOTECH BOOKS
4762-63/23, Ansari Road, Darya Ganj,
New Delhi-110002 (India)
Phone: 23262132
E-mail: biotechbooks@yahoo.co.in

Laser Typesetting by : Deep Graphics, Delhi

PRINTED IN INDIA

CONTENTS

CONTENTS

PREFACE

The earth looks like a blue planet, two-thirds of which is covered by water and looks like the planet of water-from space. The world's lakes and rivers are probably the planet's most important freshwater resources. But the amount of fresh water constitutes only 2.53% of the earth's water. On the earth's surface it inhabits large number of species. Freshwater systems are created by water that enters the terrestrial environment as precipitation and flows both above and below ground towards the sea. These systems encompass a wide range of habitats, including rivers, lakes, and wetlands, and the riparian zones associated with them.

India by virtue of its geography, varied terrain, and climate is blessed with numerous rivers and streams that support a rich diversity of inland and coastal wetland habitats. Major river systems in the north are Ganga, Yamuna and Brahmaputra (perennial rivers from the Himalayas) and in the south, Krishna, Godavari, and Cauvery (not perennial, as they are mainly rain-fed).

India's fourteen major, 55 minor and several hundred small rivers receive millions of liters of sewage, industrial and agricultural wastes. Most of these rivers have been degraded to sewage flowing drains. There are serious water quality problems in the towns and villages using these rivers as a source of their water.

Through this book we have tried to determine the present status of River Henwal which is a very important tributary of River Ganga. The present book is written to highlight importance of aquatic ecosystems of Himalayas with special reference to Garhwal region. We wish that this book will help the Researchers, Scientists and Students those who are doing work in the field of fresh water ecology.

CHAPTER-I

INTRODUCTION

From ancient times water has played an important role in the survival of human beings on earth, whether for drinking purpose, for transportation or in irrigation. In every aspect, water is present in the base of the human civilization. Being a natural resource it has been a primary concern of man throughout his existence. Human history cannot be written without showing the interactions and interrelations between human and water (Biswas, 1970).

In chemical terms it is the mono-oxide of hydrogen, water dissociates into (HOHààH^{+}+ OH^{-}) due to its dielectric constants, salts get easily dissolved in water than in any other solvents. Thus water is also known as Universal Solvent. The natural water contains many minerals and salts dissolved in it. Fresh water resources are finite and will be vanished if not been taken care. If we see earth through moon we found most of the part is water. Around 71% of the earth covers with water. Out of total water present on earth only 3 percent is fresh water present in different forms like river, icecaps, glaciers, underground reservoirs etc. The water which we use is 1% (approx.) which is present in rivers, reservoirs etc. rest is present in the ice form. As reported by Deming (1957), if the total ice cover of the world were to melt slowly, all the rivers of the World would keep flowing for 830 years. But the degrading environment and the devastating climatic changes due to ozone hole, greenhouse effect and global warming etc. have endangered the very existence of life. It is necessary to study and evaluated the present status of the water bodies to identify the undisturbed or less degraded hill streams and the factors responsible for the degradation. The quality/potability of water that is consumed defines the base line of protection against many diseases and infections. Prevention at the source *i.e.* end of the pipe is almost invariably the best method of ensuring safe drinking water and to be preferred to treat a contaminated water supply to render its suitability for consumption (WHO, 1985).

In India, 14 major river systems shares about 83% of the drainage basin. The Himalayas are the source of fresh water rivers so they can be called "the cradle of major rivers" that supports the life of many ecosystems. The Himalayan rivers receive about 20-30 % of their water from melted snow during the summer and monsoon seasons. The Himalayan region is also important in geo-politically context which has been realized and attracted the attention of country's planners towards the developmental activities of this mountain region.

Due to orogenic movements during tertiary period the formation of Himalayas was took place. The Himalayas are not only protector of our country but also play a key role in the social, cultural, and economic life of the people of Indian continent. This great Himalayan mountain system, which spreads from north-west to north-east, engulfs wholly or partially, 12 States and Union Territories. Its special geo-physical and eco-biological features, along with its significantly supportive aspects towards the life system of the entire north plains provide a mini world by itself.

The entire region is endowed with a wide range of physio-graphic, climate, soil and biological wealth. It has wide altitudinal range (3000-8000m) with rich diversity of habitats providing a variety of microclimate and niches not only for plants but also for animals. The vegetation, along an altitudinal gradient comprises of sub-tropical, temperate, sub-alpine and alpine types and supports a great variety of forests with unique species that vary from east to west and from low to high altitude. These forests have traditionally played a key role in safeguarding the ecology and environment of the subcontinent.

Garhwal Himalayas lying between the latitude of 29° 26' - 31° 28' N and longitude 77° 99' - 80° 6' E with an area of 30,000 km (approx.). In the Garhwal Himalayan region a vast source of fresh water bodies are present in the form of rivers, rivulets, tributaries and springs fulfilling the freshwater requirement of the population of Garhwal region. The major rivers of this region are Bhagirathi, Yamuna, Alaknanda, Bhilangana, Mandakini, Pindar, etc. with having snowy cold water with very high current.

One of the important river system flowing through Garhwal Himalaya is Ganga river system. Ganga river system has five source rivers, *viz.* the Bhagirathi, Mandakini, Alaknanda, Dhauliganga and the Pindar. All these rivers jointly form the Ganga River which emerges from mountains at Rishikesh and flows down to Haridwar and to the northern plains of India.

If we look globally all the water resources are over strained. Sewage and agricultural runoff, discharge of industrial effluents and other chemical in the rivers and lakes has poisoned the water bodies and decline the quality of water. Due to continuous discharge of chemicals, wastes and nutrients abiotic and biotic factors have been severely affected and today if we compare the data from five years back study we'll found a large difference which shows how badly these freshwater bodies are affected. According to WHO (1985), the quality/portability of water that is consumed defines the base line of protection against many diseases and infections. Prevention at the source *i.e.* end of the pipe is almost invariably the best method of ensuring safe drinking water and to be preferred to treat a contaminated water supply to render its suitability for consumption.

In a report on the 'State of the planet' by UN and the Washington Bank, freshwaters are identified to be the most threatened out of all the ecosystems in the world. The fresh water consumption has outpaced global population growth by 100% during this century. An estimated 54 % of the world's fresh water resources are reported to be in use presently. The heavy load of exploitation has given rise to a long list of challenging problems, especially in Asian countries. The scarcity, water pollution, eco-shift, ecological disbalance and extinction of biological lives have an ultimate effect on the health of man and its environment. Till today, the major basins like Indus, Ganga and the Brahmaputra serve as the Water Towers of the Himalayas. These directly or indirectly get connected by the other tributaries and also large rivers. In practical terms, we all are aware of the fact that without water there is no life, so rivers are the only life giving support. They are the backbone for sustaining life and its development. It is used for drinking, washing, irrigation and also for energy generation. The rivers have their own purifying power that is called the putrification but extensive use, misuse and over exploitation of this resource as any other natural resource has reduced this natural capacity of self-cleaning of the rivers. Water is infact one of the primary requirement of both living and non-living cycles in and around us. Water quality index has been considered as one of the most effective ways to inform our policy makers and the general public about the quality trend of water.

Instead of efforts of the non-governmental and governmental agencies/ organizations the indiscriminate discharge of the industrial, domestic sewage and other wastes from market, business centers, hotels and recreational centers are a continuous practice. This discharge has disturbed the aquatic environment to such an extent that it has hampered the very progress of the project of preserving the natural environmental quality of the Ganga River. The rich macro and micro biota existing in its ecosystem is facing a severe stress, which may result in shifting and even extinction.

Other than the daily influent of wastes, human and animal carracases, flowers, papers, clothes and plastic bags etc., and the wastewater with various characteristics from industries, domestic places and recreational sources find their ultimate way to the river. The bathing festivals like Makarshankranti, Guru Purnima, Ganga Dusshera, Vaishakhi and the Kumbh mela are the high times when thousands and lakhs of pilgrims or teerath yatrees take bath daily. All these have catalyzed the chemical as well as the biological problems.

With having a vast treasure of water in the Garhwal Himalayas, the region has a vast scope of fish and fisheries. But if we look towards the records of freshwater fisheries we found lack of information available due to poorly studied. The data which is available is also from some of the common location due to which there distribution, dynamics etc. features of freshwater are still to be studied. Thoughout the world approximately 22,000 species were recorded out of which 2,550 (12% approx.) are recorded in India. As the population and pollution level is increasing and the quality of freshwater in water bodies is declining it is essential to study, identify and classify the fishes along with there abiotic and biotic factors which helps to define the ecosystem for the sustainable fisheries in the country.

The Garhwal region is rich in fish diversity in natural conditions mainly *i.e.* in rivers lakes and reservoirs. These water sources not only fulfill the need of water for electricity generation, irrigation, household purposes and for other natural requirements but also provides a ground for coldwater fisheries (Mishra *et al.*, 2003). Due to increase in population and its needs our hill areas which are the treasure of natural resources affected severely in context of construction of dams for power generation, for extraction of mineral, mining of coal etc., which has led to increase in the natural disasters like soil erosion, land slides, flash floods and many more. As forests are the major source of fuel wood, wood for house hold purposes and for cottage industries which has very badly affected the entire region of Garhwal Himalayas. These anthropogenic factors are the main cause of loss of biodiversity, ecological disturbance, drastic change in physico-chemical and biological factors of this region and decline of water quality, finally causing the loss of aquatic environment.

Due to the deforestation along the bank side of the streams and rivers raises the water temperature and affects the occurrence and abundance of fishes (Phillips, 1929, 1931). Construction of large dams and projects like "Tehri Dam" and "Maneri Bhali Project" on the name of development has increased the deforestation in the catchment areas. Not for only these huge projects but for agriculture and horticulture practices also the cutting of trees and diversion of streams and small rivers are being going on in the Garhwal region.

The degradation of environmental assets, loss of biodiversity and decline of water level and water quality are brought by increase in technology which has severely affected the fish species by blocking there movements from one region to other for their seasonal migration and for breeding purposes. The human activities like water management practice, pollution, cutting of forests, establishment of small and large scale industries and urbanization has disturbed the ecological balance.

The effluent from various sources contains huge amount of many organic and inorganic chemicals along with the heavy metals. This has a devastating toxic effect both individually and synergistically. Not only do these alter the physico-chemical characteristics, the decrease in the dissolved oxygen (DO), increase in bio-chemical oxygen demand (BOD) and chemical oxygen demand (COD), alkalinity, other organic constituents and quantity of heavy metals but also affect the biological life prevailing in that aquatic environment. The biota present and its density are given significant importance in any reservoir or limnological study.

Temperature, dissolved oxygen (DO) and organic matter are some of the important factors that control the growth of aquatic life. The studies performed in the past reflect the presence of *Escherichia coli, Salmonela* sp. and many other such pathogenic bacteria which are prominently found in fecal matter to be also present in the water of Ganga (Bonde, 1977). Phytoplankton is supposed to be the center of aquatic food web. These serve as the food for both upper & lower strata in the food triangle. Changes in the zooplankton community structure and the predators that feed on zooplankton often add due to anthropogenic acidification of lakes. Zooplankton is microscopic free-swimming animalcule of any aquatic system. These are also the primary consumers of the phytoplankton. These serve as the major food for the fishes.

Organisms like phytoplankton belonging to primary producers are key indicator of trophic phase of water body and play a vital role in aquatic ecosystems. Phytoplankton are small aquatic organisms whose position (horizontal and vertical) in a body of water is determined primarily by fluid motion (Guillard and Ryther, 1962). The range in size from less than 1 μm diameter bacterial cells to larval fish centimeters long. They can span virtually the whole food chain, from primary producers gaining energy from chemosynthesis and photosynthesis. Their numbers are so large, and their distribution so widespread, that the solar radiation reacted from the river surface has reduced strength at the peak absorbance of the photosynthetic pigments that the phytoplankton contains. With such a large and dynamic biomass, the phytoplankton is a major component of global biogeochemical cycles (such as carbon, nitrogen etc) (Hurtt and Armstrong, 1996). They are widespread in marine as well as fresh waters and also occur on rocks or soil.

Phytoplankton abundance is controlled by several physico-chemical factors of the water system. The dominant genera in algal groupings change not only spatially but seasonally as physical, chemical and biological conditions changes in the water body. Changes like these indicate that the ecosystem and its associated organisms are under stress or that the ecosystem has become unbalanced. As a result, there could be possible implications for the intended uses of the water and even possible risks to human health. The responses of biological communities, or of the individual organisms, can be monitored in a variety of ways to indicate effects on the ecosystem. The co-existence and abundance of certain species at particular locations can indicate, for example, whether that habitat has been adversely altered. The reactions of individual organisms, such as behavioral, physiological or morphological changes, can also be studied as responses to stress or adverse stimuli (for example, caused by the presence of contaminants).

Uttaranchal, a newly created hill state of India, is enriched with aquatic ecosystem of various disciplines like rivers, streams, lakes and reservoirs. The rivers are snow fed rivers and spring fed rivers. There are large number of snow fed rivers and streams such as Alaknanda, Bhagirathi, Mandakini and Ganga. Besides the snow fed rivers, there are so many spring fed rivers such as Bhilangna, Nayar, Gular, Song, Suswa, Henwal and hundreds of rivulets. They all contain a very rich and colourful fish fauna.

River Henwal is spring fed river. It originates from Kaddukhal (Near Surkanda Devi Temple, Chamba, Uttaranchal) and its length is near about 38 kilometers. This river can be divided in three different ecological sections, *viz.*, the upper, the middle and the lower section. River substratum is stony, rocky and pebbly and finally it is sandy, when it meets the river Ganga. Due to the great variation in the velocity and temperature of the water, the biodiversity may also vary.

River Henwal is the tributary of the river Ganga. Local people use the water of river Henwal in agriculture. This water is not used for drinking because the unwanted product is added in it by the villagers in different ways like bathing, sewage discharge, dumping of domestic wastes etc. Due to these activities, the physico-chemical properties of this river water may be changed. The biotic communities which are found in river Henwal are also affected by these changes.

The present study of limnological characters and their relation to the biodiversity of spring fed river Henwal a tributary of river Ganga. Henwal River is originating from Surkanda hills (2900 msl) in the greater Himalayas and merges into river Ganga at Shivpuri. The present study has been carried out by considering the importance of aquatic ecosystems of Himalayas with special reference to Garhwal region.

CHAPTER-2

SAMPLING SITES

Water samples were taken from three different sites of Henwal River i.e. from Nagni, Jajal and Kharee. This river is the tributary of Holy River Ganga proving water for irrigation, domestic and for other purposes. (Fig. 2.1)

Sampling Stations:

The following three sites were selected for the collection of water samples.

A. Jajal: This site is at a distance of 3 km from sampling site - I. This point is used for bathing and washing purpose. (Fig. 2.2).

B. Kharee: Throughout the year fecal matter, decaying leaves and flowers, wooden parts, waste clothes, food material and other local waste material can be observed at this study site. The bottom of the river at this sampling site is pebbly, covered with sand and algal vegetation (Fig. 2.3).

C. Nagni: Here the riverbed consists of pebbles, sand and stones. The basic source of pollution is the domestic effluent that is discharged into the river. The sampling was done at the left bank of the river (Fig. 2.4).

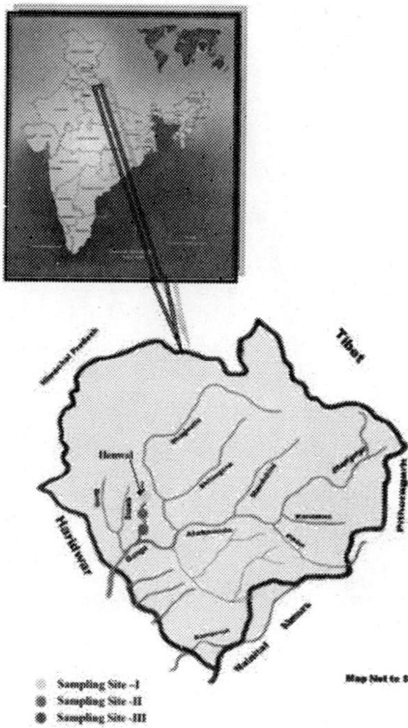

Fig. 2.1 Showing Map of River Henwal with different sampline sites

Fig. 2.2 Photograph of Jajal (Sampling Site-I)

Fig. 2.3 Photograph of Kharee (Sampling Site-II)

Fig. 2.4 Photograph of Nagni (Sampling Site-III)

CHAPTER-3

LITERATURE CITED

Today the world is still on the verge of facing environmental disasters varying environmental crisis. Those days may be repeated again when the environmental tragedies like Bhopal gas leakage and Chernobyl radioactivity leakage took place causing severe damage to human life Gupta (1996).

Our environment includes air, water, land, flora, fauna and other biota. It is like a huge fabric engulfing the whole earth in which various continuous self-generating cycles are going on. However, man, in his quest for comfort, well and power has been ignoring nature's law and broken out of these circles. Grandiose achievements of man in the field of science and technology have led to an upsurge of industrialization, urbanization growth of population, proliferation of consumer goods, dense and other big projects, which are considered to be the hallmarks of progress and development (Rana, 2000).

Reviewing of past period shows that beginning of civilization always started or born up only near to the water sources. It has been stated in Yajurveda that "anything we try to identify requires water and depends parasitically on water" to this unique indispensable part of our life we pray. Fresh water is a natural resource of fundamental importance. No life is possible without water on the earth in universe. In many respects the properties of water are unique. They seem to be especially designed for the living organism. No other liquid can replace it.

Total annual river flow of the world is 37000 cubic km. while that of Indian river is 1645 cubic km, i.e. 4.5% of total world's river water. Land mass of our country criss-cross by various rivers. In our culture and tradition rivers are not only geo-morphological and hydrological objects but have got very important place and most of them are considered sacred. They are the lifeline of many cities, towns and villages. River water is used in many ways like bathing, drinking, municipal water supply, navigation, irrigation, fishing and in industrialization. Simultaneously it is used as a course of discharge of industrial effluents, municipal sewage and disposal of solid waste.

The people of ancient times seemed to be concerned about the water quality, as they had realized that it was one of the most absolute necessities for life. A close analysis of literature in India context explains that there is a good quality and quantity of valuable data is available of lentic system in comparison to that of streams and rivers owing to the difficulties in the methodologies to be adopted in river systems. Although since along time investigations have been carried out on the ecology of rivers in respect of various physico-chemical and microbial characteristics. However, it is difficult to give the name of Scientists, Limnologists and Hydro-Biologists who worked on the several aspects of water quality and aquatic ecology. Many researchers/workers from India and abroad discussed the different kinds of aspects of water quality in terms of various characteristics. This review includes some of the important contributions made from various parts of the globe.

Many workers have studied the physico-chemical properties of water. Needham and Needham (1972) presented a guide to fresh water biology. Various studies has been made by Badola and Singh (1981), Dobriyal and Singh (1981), Singh *et al.* (1982), Nautiyal and Lal (1982 a,b), Dobriyal *et al.* (1983), Pokhriyal *et al.* (1983) & Bilgrami and Duttamunshi (1985) on the river Ganga; Dobriyal (1985), Sangu and Sharma (1985) made studies on river Yamuna and Bisht and Das (1985) presented study on abiotic parameters and aquatic insects of two cold water lakes in Kumaon. Kudesia and Verma (1985) performed a study of industrial pollution on Kali River. Nautiyal *et al.* (1986) studies the quality and potability of water of Alaknanda River and gave a note on bacterial density as an indicator of pollution Dobriyal and Singh (1987) gave a study on Rhitho plankton in Garhwal hill streams. Chopra *et al.* (1994), studied the effect of domestic sewage on self-putrification of Ganga water at Rishikesh. Hamilton (1822) made an account of the fishes found in the river Ganga and its branches. Forel (1892) can be undoubtedly considered as the founder of modern Limnology. He was the first to study the chemical, physical and biological characteristics of the lakes. Hora (1921) made his study on some new or rare species of fishes from Eastern Himalayas.

The studies on river system continued, Velz (1947) explains the process of recovery and self-purification of river water. Abdin (1948 a, b) studied the relationship of physico-chemical factors to algal growth and seasonal distribution of plankton and sessile algae in river Nile. Chacko and Ganapati (1949) made hydrobiological observations on the river Adyar. Further more, Chacko and Ganapati (1952) also studied the hydrobiology of Surli river of Madurai District to determine its suitability from the introduction of the rainbow fish trout. Ahmed (1952) in Pakistan reported various stocking combinations of local major carps based on the ecological conditions of pond water. Berner (1951) investigated the limnology of lower Missouri river Das and Moitra (1955) said that fish belonging to surface, column and bottom feeding categories have been sub-grouped according to food character they consume, into herbivorous, carnivorous and omnivorous. Das and Movtra (1955) studied the feeding habits of fresh water fishes of

Uttar Pradesh. Griffith (1955) analyzed the plankton yield in relation to certain physico-chemical parameters in Lake Michigan. Motwani *et al.* (1956) studied about the river Sone. Hora (1957) made contribution in the study of physiology of fishes. Oliff (1959) analyzed the hydrobiology of Tugela River. Holden and Green (1960) studied the hydrology of plankton of river Sokoto. Mahajan (1961) studied the fish fauna of Muzzaffarnagar district, Uttar Pradesh. Reid (1961) investigated the ecology of inland water and estuaries at New York. Chugunova (1963) studied the growth and age in fishes, Ruttner (1963), Mackereth (1963), Egborge (1986), Bachmann (2001), Cooke and Kennedy (2001), Egborge (1979), Brune (1953), Carlson (1977), Canfield and Bachmann (1981), Egborge (1978c) have studied the physicochemical parameters of various freshwater bodies in context of physicochemical, biological parameters. Eddy (1934) made a study on freshwater planktonic communities. Ruttner (1963) in fundamentals of limnology has excessively deserved Limnobiotic components. Moitra and Bhowmik (1968) studied the fresh water fisheries pond in Kalyan, west Bengal.

Ward and Seibert (1963) studied the limnology of Dow Lake. Weible *et al.* (1964) declared that urban land run off also contribute to pollution of river water. Deshmukh *et al.* (1964) made studies on the water of river Kanhan (Nagpur) Ray *et al.* (1966) studied some aspects of the river Ganga and Jamuna at Allahabad (U.P.). Mills *et al.* (1966) observed the effect of human activities on fish of river Illinois river. Ward and Whipple (1959) made further contributions in the study of the fresh water biology. Train (1967) evaluated the quality criteria for water. Venkateswarlu and Jayanti (1968) studied the hydrobiology of the river Sabarmati to evaluate the water quality. Ghosh and Basu (1968) while working in Hooghly estuary reported that the waste discharges from a chemical factory contains both inorganic and organic pollutants, which highly effects the downstream of the discharge point as compared to upstream point. Due to high dilution and heavy mixing the estuary recovers from the hazards of pollution, as it travels at a distance of about 3.5 km downstream. Muller and Forstner (1968b) studied the general relationship between the water discharge and suspended sediment concentration. Verma and Shukla (1969) observed the pollution in perennial stream Khala, by the sugar factory effluent near Laksar (Dist. Saharanpur). Das and Akhtar (1970) studied the microorganisms of Dal Lake, Kashmir. Hynes (1970) also observed the biology of polluted water. Tan (1971) studied the occurrence and ecological adaptation of zooplankton in Tansh river of Taiwan. Grove (1972) investigated some south west African rivers with respect to dissolved and suspended load. Prescott (1972), Hart and Fuller (1974), Macan (1974) were among those who gave the main method for the analysis of water samples for physical, chemical and biological parameters. Cosy and Newton (1973) studied the chemical compounds and flow of the river Fromes and its tributaries. Stumm and Bilinski (1973) studied trace metals in natural water. Sehgal (1973) studied fishes and fisheries of Aligarh with notes on fish catches.

In context of hill streams, also studies have been carried out by various workers. Schoener (1974), Bustard and Narver (1975), Bovee and Milhous (1978), Hamilton and Bergersen (1984), Sullivan (1986). Beschta and Plats (1987) Bisson *et al.,* (1988), Statzner *et al.,* (1988), Davis and Barmuta (1989), Bilby and Ward (1989), Ward (1989), Heede and Rinne (1990), Sedall *et al.,* (1990), Hughes *et al.,* (1990), Copp (1991), Baker *et al.,* (1991), Schlosser (1991), Kreshner and Snider (1992), Vadas (1992), Cuffney *et al.,* (1993), Hawkins *et al.,* (1993). Vadas (1994), Flosi and Reynolds (1994), Rankin (1995), Montgomery *et al.,* (1995), Moore *et al.,* (1995), Bailey (1995), Omernik (1995), Bison and Montgomery (1996), Barbour *et al.,* (1996), Parsons and Norris (1996), Vadas and Orth (1998), Rabeni (2000), Keim and Skaugset (2002), Johal *et al.,* (2002) and Benson *et al.,* (2005).

The limnological studies on temperate water were contributed by Chandler (1944), Thienman (1922), Pearsall (1930), Lindman (1942), Pennak (1949). They laid down firm foundation and provided guidelines and inspiration to the present day workers. Detailed investigations on the relationship between productivity and other ecological parameters were carried out later by Goldman and Wetzel (1963), Prati *et al.* (1977), Singh (1982), Lund (1972), Jenkin (1942) and Edmondson (1977) etc.

In India limnological aspects of several fresh water ecosystems have been studied by David *et al.,* (1969), George (1976), Sharma (1978), Khan and Jutshi (1970), Unni (1982), Rao (1988), Belsare (1990), Welch (1948), Imvebore (1964), Khanna (1993), Arvind Kumar (1995), Sindh (1997), Phillips (1927), Monizsette *et al.* (1978), Hutchinson (1967), Pathak, S.K. and Mudgal (2005).

Badola and Pant (1973) made a study on the fauna of the Garhwal Hills. Jhingran (1975) made study on the fish and fisheries of India. Verma and Dalela (1975a) performed study on river pollution. Robinson and Kallar (1976) compared the water characteristics of four major northern west rivers of Virginia Dabron (1976) observed physical and chemical factors of a vernal temporary pond in western Canada. Bowakar (1976) investigated the limnology of Mwenda Mouth Lake, Korea. Agarwal *et al.* (1976) monitored the physico-chemical characteristics of river Ganga at Varanasi. Bonde (1977) observed *Escherichia coli* as bacterial indicator of water pollution. He further reported *Salmonela* sp. and other pathogenic bacteria found in fresh water. Goyal *et al.* (1977) also noted occurrence and distribution of bacterial indicators and pathogens in canal communities. Cook (1977) monitored the environmental pollution by heavy metals. Patrick (1978) used algae as indicators of pollution to observe trace metals in water body. Jhingran and Sehgal (1978) studied the cold water fisheries of India. Venkateshwarlu and Menon (1979) gave a list of fishes of river Ganga and its branches. Badola (1979) pioneered limnological studies on rivers of Garhwal Himalayas. Govindan and Sundaresan (1979) studied the seasonal succession of algae in relation to pollution levels in Adyar river. Bharti *et al.* (1979) noticed physiological imbalance in fresh water algae due to hexavalent chromium metal. Bilgrami and Dutta Munshi (1979) studied the limnology of river Ganga in relation to human activities from Barauni – Farrakka. APHA, AWWA and WPCF (1989) produced the standard methods for the examination of water and wastewater. Mahatre *et al.* (1980) studied the effect of industrial pollution on aquatic system of Kalu river. Badola and Singh (1981) studied the fish and fisheries of river Alaknanda. Mohanty (1981) studied the water quality of some water bodies of Bhubneshwar.

Horned (1982) investigated the water quality of Neuse river in terms of variability, pollution loads and long-term trends. Jeejabai and Rajendra (1982) observed the changing pattern of plankton composition as indicators of water pollution. Prasad and Yashpal (1982) emphasized on diatoms as indicators of pollution. Daniel *et al.* (1982) analyzed water quality of French Broad river, based on water collected at Marshall. Raina *et al.* (1984) made pollution studies on river Jhelum. Somashekhar (1984) performed study on the river Cauvery. Chattopadhya *et al.* (1984) made pollution studies on the pollution status of river Ganga at Kanpur. Verma *et al.* (1984) observed the pollution and saprobic status of eastern Kalinadi. Trivedi and Goel (1984) presented a comprehensive work on chemical and biological methods for water pollution studies. Bhatt *et al.* (1985) studied the hydrobiology and phytoplankton population in river Kosi. Holander (1986) did microanalysis of scales of poecillid fishes and reported the teeth like structures called lepidonts Sharma (1986) published a report on the effect of physico-chemical factors and benthic fauna of river Bhagirathi, Garhwal Himalayas. Aggarwal (1986) studied ecology of Chambal river at Kota, and Singh *et al.* (1983) presented geographical distributional list of icthyofauna of Garhwal Himalayas with some new records. Reddy and Venkateswarlu (1987) on the river Tungabhadra and Sivakumar *et al.* (1987) on the river Amravati provided valuable additional information on physico-chemical parameters. Dobriyal and Singh (1988) presented the ecological basis for icthyofaunal variations in two hill streams of Garhwal Himalaya. Nautiyal and Singh (1989) made observations on the river projects and the endangered Himalayan *Tor putitora* in the uplands of Garhwal region. Gautam *et al.* (1989) observed the seasonal variation of certain oxidation-reduction characteristics of the river Bhagirathi. Gautam (1990) studied the sediments of the river Bhagirathi, Garhwal Himalaya. Nautiyal (1990) explained the ecology of the Ganga river system in the upland of Garhwal Himalaya environment Khanna and Badola (1990) studied the icthyofauna of the river Ganga at foothills of Garhwal Himalayas. Borego *et al.* (1990) reported *Cephalophage* as an indicator of pollution. Moyle and Williams (1990) studied the biodiversity loss in temperate zone and reported the reasons for decline of native fish fauna in California and suggested methods for conservation.

Joshi *et al.* (1991) performed a study on the planktonic and benthic components of three selected tributaries of river Ganga between Devprayag and Rishikesh. Pathak (1991) studied some aspects of ecology of the rivers Sarju and Gomti of the Kumaun Himalaya. Joshi and Pathak (1991) performed a relative study of some physico-chemical parameters of sewage water at Uttarkashi. Singh (1991) published a book "Advances in limnology". Joshi *et al.* (1991) made an attempt to study diurnal variation in physico-chemical characteristics of Alaknanda river at Srinagar during winter season. Tombi (1992) studied the ecobiology of hill stream fishes of Manipur with special reference to culturable possibilities. Ramesh *et al.* (1992) evaluated the water quality in three rivers of Andhra Pradesh. Gautam (1992) suggested some considerations on stone removals from river basin and resultant effects. Mazhar and Kapoor (1992) made a limnological study on Doraniu River at Bareilly.

Several studies have been conducted on the substrate classification and its importance to biotic communities by several workers (Wentworth (1922)), Cummins (1962), Postma (1967), Rhoada and Young (1970), Williams and Sale, (1981), Shulman (1984), Allen (1985), Bain *et al.*, (1985), Day *et al.*, (1989), Jones (1991), Burke (1991), Levin (1991), Stein *et al.*, (1992), Murie *et al.*, (1994), McGehee (1994), Eklov and Diel (1994), Macpherson (1994), Keefe and Able (1994), Able *et al.*, (1995), Auster *et al.*, (1995), Langton et al., (1995), Szedlmayer and Able (1996), Szedlmayer and Howe (1997), Armantrout (1999), Bridge (2003), Merilaita (2003), Webster and Hart (2004), and Johal and Rawal (2005).

Attention has been focused on the relationship between the characteristics stream water quality and their correlation with various biota by several researchers (Hynes (1960), Kuehne (1962), Harrel and Dorris (1968), Bomann *et al.*, (1969), Likens *et al.*, (1970), Huet (1972), Thomas and Crutchfield (1974), Karr and Dudley (1981), Omernick *et al.*, (1981), Boyd (1982), Lawrence *et al.*, (1985), Nautiyal (1986, 1990), Meyer *et al.*, (1988), Osborne and Wiley (1988), King (1988), Close and Davis-Colley (1990), Abohweyere (1990), Gregary *et al.*, (1991), Kemdirin and Ejike (1992), Richards and Host (1994), Richards *et al.*, (1996), Johnson *et al.*, (1997), Singh and Singh (2000), Johal *et al.*, (2001), Fafioye *et al.*, (2005), Sharma *et al.*, (2007).

A considerable amount of work has been done on the phytoplankton density and their correlation with the environmental factors throughout the world (Eddy (1934), Chandler (1940), Ganpati (1940), Ganpati and Chacko (1951), Das and Srivastava (1956), Das (1959), King (1970) John (1971), Young and King (1973), Shapiro (1973), Nautiyal and Lal (1978), Lamb and Lowe

(1981), Fisher *et al.*, (1982), Traan and Lindstrom (1983), Nautital (1984), Watanbe *et al.*, (1986), Stadecek (1986), Welch *et al.*, (1988), Nautiyal (1986, 1990), Steinman *et al.*, (1989), Horner *et al.*, (1990), Iversen *et al.*, (1991), Singh *et al.*, (1994), Allan (1995), Wasson *et al.*, (1995), Fjellheim and Raddum (1995), Hoffman (1996), Lamberti (1996), Stevenson (1996), Sobezak and Burton (1996), Nautiyal *et al.*, (1996), Nautiyal (1996), Joshi *et al.*, (1996), Badoni *et al.*, (1997), Flemer and Livingston (1998), Jack *et al.*, (1996), Vavilova and Lewis (1999), Biggs *et al.*, (1999), Ramanibai and Elayaraja (2001), Koteswari and Ramanibai (2004) and Negi *et al.*, (2007), In Indian fresh water, planktonic information have been given by Chacko and Sreenivasan (1955), Chakraborthy *et al.* (1959), David (1969), Zafar (1966), Rao (1987), Adoni (1985), Rao *et al.* (1988), Kumar (1995), Khanna *et al.* (1999), Suvarna and Somshekhar (2000), Khanna *et al.* (2000). Primary productivity has been considered as a major study by some workers like Sreenivasan (1965), Hussaini (1967), Lewis (1974), Dutta (1997), Sharma (1976), Singh and Desai (1980), Khan and Zutshi (1980), Gautam (1988), Ahmed (1990), Prasad (1990), Singh (1996).

Khanna *et al.* (1997) investigated population of green algae in relation to physico-chemical factors of river Ganga. Kaur and Khulbe (1998) studied occurrence, distribution and ecology of water Molds in Nanak Sagar reservoir. Khanna *et al.* (1998) studied the phytoplanktonic communities in relation to certain physico-chemical parameters of Ganga. Singh and Ahmed (1990) worked on comparative study of the phytoplankton of river Ganga and pond of Patna. Khanna *et al.* (1999) studied correlation between abiotic factors and planktonic population in river Ganga at Rishikesh (U.P). Khanna and Singh (1999) observed diurnal variation in microbial ecology of Narkewala pond at Ferupur, Hardwar. Khanna *et al.* (1999) studied fishes and their ecology of the river Suswa at Raiwala, Dehradun. Badola and Badola (1999) gave observations on destructive fishing methods and recommendations to protect fish fauna in Garhwal Himalayas. Chugh (2000) studied the seasonal variation in microbial ecology of river Ganga at Hardwar. Khanna *et al.* (2000) observed 20 genera of phytoplanktonic population of a pond they revealed that highest phytoplanktonic population was recorded in winter season while lowest in monsoon season. Khanna *et al.* (2001) studied fluctuation in the population density of macro-Invertebrates of river Ganga at Pashulok Barrage, Rishikesh. Khanna and Singh (2002) published a book on pond fish ecology and economics. Basu and Mitra (2002) studied about pollution of river Damodar at Burdwan with special reference to metals, COD and pH. Kumar and Sharma (2002) studied the water quality of river Krishni (Part-II Biological characteristics and bio indicators). Kala and Sharma (2002) worked on effect of physico-chemical factors on phytoplankton population.

A significant amount of work in different parts of the world on qualitative and quantitative variations in zooplankton abundance has been carried our by various workers George (1958), Sladecek (1973), Cairns (1974), John (1971), Nautiyal and Lal (1978), Meyer and Johnson (1983), Singh and Singh (1985), Bilgrami and Dutta Munshi (1985), Prat (1987), Laal (1988), Haque (1988), Foissner (1988), Griffith and Perry (1991), Corkum (1992), Tuchman and King (1993), Richard and Host (1994), Smock (1994), Koetsier *et al.*, (1996), Edwards and Huryn (1996), Hawkins *et al.*, (1997), Davidson *et al.*, 1998), Kobayashi *et al.*, (1988), Pozo *et al.*, (1998), Vinogrador *et al.*, 1998), Akopian *et al.*, (1999), Biggs *et al.*, (1999), Barbara and Higler (2001), Bretshko and Waidbachor (2001), Wootton and Gee (2001), Sandreson *et al.*, (2003), Fuma *et al.*, (2003), Koteswari and Ramanibai (2003), Zafer and Sultana (2005), Negi *et al.*, (2007).

Singh *et al.*, (2002) observed the distribution and seasonal abundance of benthic macro invertebrates in the Bisar sarovar of Gaya, Bihar. Pathani *et al.* (2002) described some physico-chemical parameters and primary productivity of river west Ramganga. Bahuguna and Badoni (2002) made a study on qualitative and quantitative productivity of fish food from three important tributaries of river Alaknanda, Garhwal· Dutta *et al.* (2002) presented an overview of the fish diversity of Ranchi district. Fokmare and Musaddique (2002) investigated the physico-chemical characteristics of Kapashi lake and Purna river waters in Alok district of Maharashtra. Sukumaran (2002) made observation on coliform bacteria an index of aquatic pollution. Ahmed and Begum (2002) monitored the bacteriological quality of water in hill stream of Khanpara during different seasons.

The occurrence of benthic life in freshwater ecosystem has a direct effect on the biotic and abiotic environment in which they inhabit. The relationship between macro-invertebrate biotic and abiotic factors have been the subject of deep study, Gaufin and Farzwell (1952), Chakrabarty *et al.*, (1959), Hynes (1960), Keup *et al.*, (1966), Wilhm and Dorris (1968), Harrel and Dorris (1968), Hynes (1970), Minshall and Minshall (1977), Connell (1978), Riece (1977), Bass and Harel (1980), Donald *et al.*, (1982), Fisher *et al.*, (1982), Michael (1984), Depiereux *et al.*, (1984), Sousa (1984), Reice (1985), Omerod (1987), Omerod and Edwards (1987), Resh *et al.*, (1988), Cimmins (1992), Richard and Minshall (1992), Duffied and Nelson (1993), Gower *et al.*, (1994), Bass (1994), Richards and Host (1994), Resh (1995), Deshon (1995), Pires *et al.*, (2000) Boyero and Bosch (2004), Soininen and Kononen (2004).

Several studies have been done on the fishes of Indian sub-continent for the betterment of commercial fisheries and for their taxonomy. Shelford (1911), Hora (1940), Huet (1949), Silas (1951), Rajan (1963), Lotrich (1973), Jayaram (1974) Schoener (1974), Kushlam (1976), Horowitz (1978), Gorman and Karr (1978), Evans and Noble (1979), Matthews and Hill (1980), Grossman *et al.*, (1982), Fraser and Cerri (1982), Finger (1982), Schlosser (1982 a,b), Power and Matthews (1983), Moyle and Senanayake (1984), Probst *et al.*, (1984), McClendin and Rabeni (1987), Schlosser (1987, a, b), Datta Munshi and Srivastava (1988), Arunachalam *et al.*, (1988), Gorman (1988 a,b), Kortmulder *et al.*, (1990), Fausch *et al.*, (1990), Sehgal (1990), Gregary *et al.*, (1991) Falwar and Jhingran (1991), Karr (1991), Ratherford *et al.*, (1992), Pusey *et al.*, (1993), Edds (1993), Wootton and power (1993), Cairns *et al.*, (1993), Paller (1994), Harris (1995), Arunachalam *et al.*, (1997a,b), Braaten and Berry (1997), Leveque (1997), Scott and Hall (1997), Bowen *et al.*, (1998), Sehgal (1999), Jayaram (1999), Arunachalam (2000), Johal *et al.*, (2001), Mazzoni and Iglesias Rios (2002), Katano *et al.*, (2003), Bhat (2003), Wright and Flecker (2004), Katano *et al.*, (2006).

Khanna *et al.* (1998) performed an extensive study on the fish fauna of river Ganga at Rishikesh. They reported that during their course of study various species of fishes as *Tor tor*, *Tor putitora* and *Raimas bola* were found endangered and the main reason was declared to be the polluted habitats, land slides and soil erosions besides this, the destructive methods of fishing in the river Ganga is also the prime factor of depleting fish fauna. Kaur *et al.* (1999) made an attempt to study the aquatic diversity of Kalanji Lake, Punjab. Badola and Badola (1999) gave their observations on destructive fishing methods and recommendations to protect fish fauna in Garhwal Himalayas. Kumar and Bohra (1999) observed Gastropods as indicators of pollution status of some wetlands in Santal pargana, Bihar. Khanna *et al.* (1999) studied the fish and their ecology in river Suswa at Raiwala. Kumar and Mukherjee (2000) observed the changes in mammalian and fish fauna of Doon Valley and surrounding area. Karuppasamy (2000) studied the short and long term effect of Phenyl mercuric acetate on the protein metabolism in *Channa punctatus* (Bloch). Dua and Johal (2000) used scales of *Channa punctatus* (Bloch) as an indicator for pesticide pollution and declared significant impairment upon exposure to this organochlorine pesticide. They said that on exposure to endosulfan there was disruption in lepidonts from the anterior side of the scales, thus result in loosening of hold of scales from the body of fishes.

Khanna and Bhutiani (2003) observed the limonological characterstics of river Ganga at Haridwar. Khanna and Bhutiani (2003) studied the limonological status of Satikund pond at Haridwar. Sivakumar *et al.*, (2003) studied the water quality of the river Ambaram palayam, Coimbatore District, Tamilnadu. Monichan and Sharma (2003) studied the distribution status of fish fauna in Etawah district, U.P. Jena *et al.*, (2003) performed studies on water quality parameters around Sagar Island, Sundarbans, Wanganeo *et al.*, (2003), revealed that lotic system under temperature climatic condition support less fish diversity in comparison to tropical climatic conditions. Sivakumar *et al.*, (2003) presented a study on some biological aspects of fresh water fish. Khanna and Bhutiani (2004), studied ecology of some fishes of river Ganga at Gohri Ghat, Garhwal in relation to abiotic factors. Esmaeili *et al.*, (2001) performed SEM studied on body scale of a tooth carp, *Aphanius ginaonis* (Holy) from the hot spring of Geno, South Iron. Joshi (2005) made study on the ecology and contribution of smaller mountain tributaries to the productivity of Major River. Bhagirathi-Ganga- in Himalayan region. Johal *et al.*, (2005) made a study on the rivers of major landscape features, representing visible portion of the aquatic system. Jindal and Singh (2005) also studied the ecology of benthos of some fresh water systems of Punjab. Khanna *et al.*, (2007) studies the biotic and abiotic factors of Song River at Dehradun. Kesre *et al.*, made a study on the physico-chemical parameters for a reservoir at Khandwa District (M.P.). Vishnoi *et al.*, also studied the abiotic status of Song River and its relation to zoo and phytoplankton at Nepali farm (Dehradun).

CHAPTER-4

MATERIALS AND METHODS

Selected physico-chemical parameters were analyzed following the standard methods of APHA.AWWA-WPCF (1998), Trivedi and Goel (1984) and Mathur (1999) while biological and microbiological parameters are studied following the standard methods of Welch (1998), Khanna (1993) and APHA.AWWA WPCF (1998).

Sampling programme and analysis

Bearing in mind the river characteristics and the water quality requirements, the following routine sampling and analysis programmes were devised: -

Number of sites	: Three
Collection of samples	: Monthly
Time of sample collection	: Between 7:00 A.M. to 11:00 A.M.
On site analysis	: Temperature, velocity, DO, Free CO_2
Analysis within 24 hours	: State variables and biological parameters

Chemicals

The chemicals used in the present investigation were of analytical grade and were supplied by E. Merck India.

Glassware

All the glasswares used were of corning grade, manufactured by Borosil India Ltd.

A. Bottles

DO bottles of 300 ml capacity for D.O. analysis and for sample collection, reagents bottles 100 to 1000 ml capacity were used of Borosil make.

B. Burettes / Pipettes

25 ml, 50 ml, 100 ml graduated burette and different volume size pipettes graduated / volumetric of borosil make were used.

C. Beakers

100 to 1000 ml capacity of borosil make used.

D. Conical Flask

Conical flask from 150 ml to 1000 ml capacity of borosil make.

E. Measuring cylinder

Graduated measuring cylinders from 25 ml to 250 ml of borosil make were used.

Others

Plastic wash bottles of 150 ml capacity plastic droppers and glass funnels, tripods, whatman filter papers, crucible, china dishes, spirit lamp, wire gauge, Bunsen burner.

Instrumentation

Following instruments were used during the analysis:

a. Thermometer - °C (Mercury thermometer)

b. pH meter - Digital pH meter Century, model - CP 901 systronics make.

c. Conductivity Meter - Micro-processor based Century, model CP 631.

d. Electronic balance - Precisa, Model 310M.

e. Hot air Oven

f. Laminar flow U.V. Chamber

g. Microscope - Suswox Optic Model 5K-88.

h. Distil Water Plant - Tanco.

i. Lab System Analyser - 7 Filter medi lab systems.

j. Rotatory Flask shaker - Tanco

Sampling methods

(A) Cleaning and sterilization of glassware

All the glassware's and jerricans were cleaned or washed with soda and chromic acid and rinsed with double distilled water 2-3 times.

(i) Sample bottles

300 ml capacity BOD bottles made of borosil were used for sample collection. They were washed with washing soda and chromic acid and rinsed with tap water followed by distilled water; the neck and stopper were wrapped with butter paper with the help of rubber band. The bottles were then sterilized in an autoclave at 15-lbs/ inch^2 pressure (121°C) for 15-25 minutes.

(ii) Pipettes

Pipettes of different volume size were washed and fitted with cotton plug at the upper end; these were then wrapped in butter paper and sterilized in an autoclave at 15-lbs/inch^2 pressure at 121°C for 15-25 minutes.

(iii) Other glass wares

Such as conical flasks, beakers, measuring cylinders etc. were washed carefully and then sterilized in hot air oven at 160-180°C for 2-3 hours.

(B) Collection of samples

(i) For determination of the physico-chemical parameters

For the estimation of total solids, turbidity, total hardness, phosphate, calcium and magnesium samples were taken from 2 feet below the surface water by keeping and opening jerricans below the surface water. Caps of cans were removed after dipping the can and also closed in the water after filling up of can.

(ii) For dissolved oxygen

Samples were collected in sterilized 300 ml capacity BOD bottles. The bottles were filled completely with sample water up to the rim and stopper was placed inside water only for avoiding any kind of air bubble inside it. DO was immediately fixed by adding 2 ml each of manganous sulphate ($MnSO_4$) and alkaline potassium iodide (KI) at sampling spot.

(iii) Sampling for biological analysis

For biological analysis of water, samples were collected with the help of plankton net attached with a collection tube at the base of net. For this a known amount of water was filtered through the plankton collection net and sample was collected inside collection tube. The sample was then transferred in sterilized tube of 50 ml capacity and preserved by adding one drop of lugol's solution or formaldehyde solution.

(C) Reagent preparation

1. Sulfuric acid

H_2SO_4 Concentrated (Sp. Gr.1.84)

2. Sodium Thiosulphate (0.025N)

Dissolve 24.82 gm of $Na_2S_2O_3.5H_2O$ in boiled distilled water to make final volume to one liter. A pellet of NaOH or 0.4 gm borax was added as stabilizer. This is 0.1 stock N solution. Dilute 0.1N stock solution 4 times with boiled distilled water to prepare 0.025N solution. Keep it in brown glass stopper bottle.

3. Alkali iodide azide solution

a) Dissolve 500 gm of NaOH or 700 gm of KOH and 150 gm of KI in distilled water to makes one liter of solution.

b) Dissolve 10 gm of NaN_2 in 40 ml of distilled water mix both (a) and (b) solution.

4. Manganous sulphate solution

Dissolve 100 gm of $MnSO_4.4H_2O$ in 200 ml of boiled distilled water and filter.

5. Starch solution

Dissolve 1 gm of starch in 100 ml of warm distilled water and add a few drops of formaldehyde solution.

6. Sodium hydroxide (0.05N)

Dissolve 40 gm of NaOH in one liter boiled CO_2 free distilled water to prepare 1N NaOH stock solution. Dilute 50 ml of 1N NaOH stock solution to 1 liter distilled water. Standardized the solution using H_2SO_4.

7. Phenolphthalein indicator

Dissolve 0.5 gm of phenolphthalein in 50ml of 95% ethanol and add 50 ml of distilled water. Add slowly 0.05N CO_2 free NaOH until solution turns faintly pink.

8. Methyl orange indicator

Dissolve 0.5 gm methyl orange in 100 ml of distilled water.

9. Buffer solution

a) Dissolve 16.9 gm of ammonium chloride (NH_4Cl) in 143 ml of concentrated ammonium hydroxide (NH_4OH).

b) Dissolve 1.179 gm of disodium EDTA and 0.78 gm of $MgSO_4.7H_2O$ in 50 ml distilled water.

Mix (a) and (b) solution and make final volume to 250 ml using distilled water.

10. Eriochrome black-T indicator

Mix 0.5 gm of EBT with 100 gm of NaCl (A.R) and grind.

11. Sodium hydroxide solution (1N)

Dissolve 40 gm of NaOH in one liter distilled water.

12. Potassium chromate (5%)

Dissolve 5 gm of K_2CrO_4 in 100 ml of distilled water.

13. Ethyl alcohol, 95%

14. Lugol's solution

Dissolve 10gm of KI and 5 ml of iodine crystals in 20 ml of distilled water and then add 50 ml distilled water in which 5 gm anhydrous sodium acetate has been dissolved.

15. Sulphanilic acid solution

Dissolve 600 gm sulphuric acid in 70 ml of hot distilled water and add 20 ml conc. HCl finally dilute it to 100 ml using distill water.

16. Ammonium molybdate solution

a) Dissolve 25.0 gm of ammonium molybdate in 175 ml distilled water.

b) Add 280 ml of conc. H_2SO_4 to 400 ml of distilled water and cool.

Mix (a) and (b) solution and dilute to 1 liter.

17. Stannous chloride ($SnCl_2$)

Dissolve 2.5 gm of stannous chloride in 100 ml glycerol by heating on water bath for rapid dissolution.

18. Standard phosphate solution

Dissolve 4.388 gm dried anhydrous potassium hydrogen phosphate K_2HPO_4 in distilled water and make up the volume to 1 liter, dilute this solution 100 times (10-1000 ml). This is standard phosphate solution containing 10 mg P/L (1 ml = 0.01 mg P).

Methods for physico-chemical parameters

Samples were analyzed for the following physico-chemical parameters.

1. Physical parameters

a. Water Velocity
b. Water Temperature
c. Conductivity
d. Turbidity
e. Total solids

2. Chemical parameters

(i) D.O.
(ii) Free CO_2
(iii) Acidity
(iv) Total Alkalinity
(v) Total Hardness Ca/Mg
(vi) Chlorides
(vii) Phosphates

1. Physical parameters

a) Water Velocity

The particular distance covered by a pre-weighted drifting float in a particular time represents the velocity of running water. The float may be of any size or shape and are previously weighted so as to get them submerged and have no wind effect.

$$\text{Velocity (m/s)} = \frac{D}{T}$$

Where,

D= Distance covered by float in meters.

T = Time taken by float to cover distance from one pole to another.

b) Water Temperature

The temperature of water was recorded with the help of ordinary mercury thermometer with Centigrade scale. The thermometer was dipped into the bank of the river at considerable depth for 5 minutes. Then the reading as $X^{\circ}C$ was recorded.

c) Conductivity

The conductivity of water was directly measured by the help of Water Quality Analyzer Model PE-132 and was recorded as Siemens/cm.

d) Turbidity

Turbidity is an expression of the optical property that causes light to be scattered and absorbed rather than transmitted with no change in direction or flux level through the sample. The turbidity of water was measured with the helps of Jackson's candle turbidity meter. It consists of three parts Candle, Holder and Calibrated glass tube.

Collected samples were shacked thoroughly in jerricans and were slowly transferred into calibrated glass tube of Jackson's candle turbidity meter until the flame of the candle disappears and light path was noted in centimeters. The reading (cm) was then converted into Jackson turbidity unit (JTU) using conversion table.

e) Total Solids

After proper mixing a 100 ml of water sample was taken to a pre dried and pre-weighed porcelain dish, which was then dried in an oven at 105° C. After drying it was transferred to a desiccator for cooling. Finally after cooling it was weighed and calculated with the formula as below:

$$\text{Total solid, (mg/l)} = \frac{(A - B) \times 1000 \times 1000}{\text{Volume of sample}}$$

Where,

A = Final weight of dish (with sample)

B = Previous weight of dish (without sample)

2. Chemical Parameters

a. Dissolved Oxygen (DO)

The DO content was estimated by iodometric titration and $MnSO_4$ and alkaline KI solution by graduated pipette to 300 ml of water sample (to avoid agitation). After allowing the precipitate to settle down 1 ml of concentrated H_2SO_4 was added to it slowly which results in liberation of iodine equivalent to DO present in water sample. This iodine was then titrated with the standardized thiosulphate solution ($Na_2S_2O_3$). Starch was used as an internal indicator.

The number of ml titrant used was recorded and the result expressed as mg/l with the aid of following formula: -

$$\text{Dissolved oxygen, (mg/l)} = \frac{(\text{ml} \times \text{N}) \text{ of titrant} \times 1000 \times 8}{V_1 - V}$$

Where,

V_1 = Volume of sample used

V = Volume of $MnSO_4$ and KI added.

b. Free CO_2

Ten drops of neutralized phenolphthalein indicator were added to water sample (known amount) and titrated by N/44 NaOH until the pink color flashed throughout the solution. The volume of titrant used was recorded and estimation of CO_2 was made by the following formula:

$$\text{Free } CO_2, \text{(mg/l)} = \frac{(\text{ml} \times \text{N of titrant}) \times 1000 \times 44}{\text{ml. of sample taken}}$$

c. Acidity

Acidity in water results in change of its quality, which may be due to strong or weak acids and also due to presence of some hydrolyzing salts. For analysis we have taken 100ml of sample in conical flask then 2-3 drops of methyl orange indicator was added. The solution turned pink then we titrated this sample with 0.05N NaOH. Pink colour changes to yellow at end point. Add few drops of phenolphthalein indicator and titrated further with NaOH until colour turns pink.

$$\text{Acidity, (mg/l)} = \frac{(V + V_1) \times \text{N of NaOH} \times 1000 \times 50}{\text{ml. of sample taken}}$$

Where,

V = Vol. of NaOH used with methyl orange.

V_1 = Vol. of NaOH used with Phenolphthalein.

d. Total Alkalinity

The neutralizing capacity of a strong acid is defined as alkalinity of water. The presence of Carbonates, Bicarbonates along with Hydroxyl ions makes H_2O alkaline, various water may also consist phosphates, silicates and borates.

A 100ml sample was taken in an Erlenmeyer flask followed by the addition of 4-6 drops of phenolphthalein indicator. When the sample turned pinkish, it was titrated by 0.1N HCl till the color disappeared. After this add 2-3 drops of methyl orange indicator.

When the sample turned yellowish, continue the titration till the yellow colour changed to pink.

$$\text{Total Acidity, (mg/l)} = \frac{(V \times N \text{ of HCl}) \times 1000 \times 50}{\text{ml. of sample taken}}$$

Where,

V = ml. of HCl with phenolphthalein and methyl orange.

e. Total Hardness

Total hardness was determined by using EDTA titrametric method. 100 ml sample was titrated by 0.01M EDTA solution after the addition of indicator (very small amount of EBT + 1 ml of ammonia buffer). End point was recorded at the time of first color change (pinkish to blue) and total hardness was then calculated by using the formula: -

$$\text{Total Hardness, (mg/l)} = \frac{\text{ml of EDTA used} \times 1000}{\text{ml. of sample taken}}$$

Hardness of water is measure of the capacity of water to precipitate soap, precipitated mainly by the calcium and magnesium ions present.

f. Calcium

100 ml of sample was taken in a flask, followed by the addition of murexide indicator and 11ml of 1N NaOH. After it the sample was titrated by standard EDTA solution till the color changed (pink color change to purple) this is end point. This is calculated by understated equation:

$$\text{Ca hardness, (mg/l)} = \frac{\text{ml of EDTA used} \times 1000}{\text{ml. of sample titrated}}$$

g. Magnesium

Magnesium was estimated by subtracting the calcium hardness from total hardness. It was calculated by using formula:-

Mg hardness (mg/l) = Total hardness – Ca hardness.

h. Chlorides

A known amount of sample was taken in an Erlenmeyer flask, followed by the addition of 1ml of potassium chromate indicator; it was then titrated with $AgNO_3$ solution. Slight change in color forms yellow to orange as considered as end point and the calculation was made according to: -

$$\text{Chloride, (mg/l)} = \frac{(\text{ml.} \times \text{N of AgNO}_3) \times 1000 \times 35.5}{\text{ml. of sample taken}}$$

i. Phosphate

Fifty ml of sample was taken in a flask followed by the addition of acid ammonium molybdate solution (2 ml) and 4-5 drops of $SnCl_2$ solution. A blue color appeared. The same procedure was run with a blank (distilled water) sample for comparison. Both (Sample and blank) were then kept on white paper, followed by the addition of standard phosphate solution to the blank drop by drop with the help of graduated pipette (1ml) until the color of blank reached at equivalent to the sample, at the end point the ml of standard phosphate used was placed in the equation to get the amount of phosphate.

$$\text{Phosphates, (mg/l)} = \text{ml. of solution used} \times 0.01 \times 20$$

3. Biological parameters

(1) Collection of plankton

Plankton samples were collected through fine plankton net. The known amount of water was filtered several times. Samples were mixed and collected after adding 4% formalin and brought to the laboratory for analysis. After centrifugation take 1 ml sample in a Sedgwick Rafter counting cell. Counting of the organisms was done by following formula.

$$N = \frac{(A \times 1000) \times C}{1}$$

Where **N** = number of plankton/litre of water.

A = average of plankton in one small chamber.

C = ml. of plankton concentrate.

l = Volume of original water filtered in litre.

(2) Macrobenthic fauna

Benthos were collected by dragnet, insect net and with the help of forceps. For preservation of samples, organisms were kept in Petri dishes having 5% formalin or finally in spirit. They are identified with the help of keys given by Ward and Whipple (1959), Needham and Needham (1972), Donal (1976) and A.P.H.A. (1980).

The number of macrobenthos were computed per square meter by the following formula:

$$N_{mb} = \frac{N}{AH} \text{ (Welch, 1948)}$$

Where, **N** = No. of macro-invertebrates / mt^2.

A = area sampled

$\mathbf{N_{mb}}$ = no. of macro-benthos / sampled area

H = no. of hauls.

The macrobenthos were collected from different sites with the help of hand picking by sieves or by means of a forcep and kept in phials for laboratory analysis.

(3) Collection of Fishes

Fishes were collected from different sites of Henwal River by the following methods -

Jaal, Triangular net, Scoop net, dragnets, Angling. Fishes were preserved in 5 percent formaline and identified with the help of keys given by Day (1978), Shrivastava (1968), Badola (1979), Jhingran (1975) and Sehgal (1978).

CHAPTER-5

PHYSICAL ANALYSIS

The physical parameters of water quality can be broken down into many topics such as turbidity and taste or odor to name a few. But one needs to also take into consideration the nature of the physical parameters of the ecosystem surrounding a water source to also understand the physical appearance of a later finished water. One of the best barometers of a water is its actual temperature in its natural ecosystem. Temperature affects sediment and microbrial growth among other source water characteristics. What are the variables that affect a waterway's temperature? Physical analysis of River Hemwal in different sampling sites is as follows:-

Sampling site I

The physical characteristics of sampling site- I are presented in Table 5. I. 1 and Figures Fig. 5. I. 1 to Fig. 5. I. 5

i. Water Velocity

The velocity of water was minimum 0.56 m/sec ± 0.35 in the February month in the year 2007 and maximum of 1.16 m/sec ± 0.82 in September month in the year 2008. The average value of velocity in the year 2007 was 0.80 m/s ± 0.20 and 0.86 m/sec. ± 0.17 in the year 2008.

ii. Water Temperature

The water temperature at this sampling site was observed minimum 13.00°C ± 0.25 in the month of December for year 2008 and maximum 28.5°C ± 0.57 in June in the year 2007. The average value in the year 2007 was found to be 20.46°C ± 4.62 and 20.25°C ± 4.38 in the year 2008.

iii. Conductivity

The conductivity of river water was observed to be minimum 96.54 Siemens/cm ± 1.24 in the February month in the year 2008 and maximum of 131.57 Siemens/cm ± 0.05 in the October month in the year 2007. The average values recorded in the year 2007 was 115.74 Siemens/cm ± 11.66 and 115.41 Siemens/cm ± 11.70 in the year 2008.

iv. Turbidity

The turbidity in water was minimum 19.69 JTU ± 0.00 in the month of February in the year 2007 and maximum of 331.76 JTU ± 13.72 in the month of September in the year 2008. The average value of turbidity in the year 2007 was 149.94 JTU ± 112.84 and 154.00 JTU ± 112.91 in the year 2008.

v. Total Solids

The total solids were minimum 51.76 mg/l ± 0.78 in February month in the year 2008 and maximum of 534.71 mg/l± 2.40 in the July month in the year 2008. The average value of total solids in the year 2007 was 236.79 mg/l ± 169.65 and 244.44 mg/l ± 177.58 in the year 2008.

Table 5. I. 1: Monthly fluctuation in Physical parameters of Sampling Site – I during 2006-2008

Months	Water velocity (m/sec)	Water temperature (°C)	Conductivity (Siemens/cm)	Turbidity (JTU)	Total solids (mg/l)
2007 (Sampling Station – I)					
January	0.64±0.00	15.0±0.01	98.54±0.00	29.40±0.04	69.94±0.36
February	0.56±0.35	16.0±0.04	96.63±0.00	19.69±0.00	54.78±13.70
March	0.60±0.36	20.0±1.26	118.17±0.03	34.78±0.08	109.23±0.25
April	0.63±0.04	22.0±0.07	121.49±0.02	77.49±7.85	134.64±0.24
May	0.63±0.21	27.0±0.53	122.18±0.01	176.49±7.81	189.40±0.52
June	0.76±1.18	28.5±0.57	120.26±0.04	209.70±37.86	210.57±0.16
July	0.99±0.00	24.0±2.08	124.62±0.26	289.76±162.00	500.14±0.39
August	1.06±0.02	23.0±0.25	126.48±0.02	329.89±3.86	476.85±7.70
September	1.11±0.59	20.0±0.00	120.34±0.00	294.46±7.00	489.74±30.19
October	0.98±0.31	19.0±0.18	131.57±0.05	184.64±13.72	337.18±6.36
November	0.86±0.11	17.0±0.00	106.10±0.00	98.67±3.86	179.17±0.26
December	0.72±0.00	14.0±0.35	102.54±0.01	54.34±1.02	89.79±0.18
Average ±Sd	0.80±0.20	20.46±4.62	115.74±11.66	149.94±112.84	236.79±169.65

contd. on next page...

contd...

2008 (Sampling Station – I)					
January	0.71±1.24	15.0±1.74	97.63±0.86	31.93±0.16	76.48±0.16
February	0.63±0.18	17.0±1.16	96.54±1.24	21.49±7.16	51.76±0.78
March	0.67±2.15	21.0±2.15	116.43±0.26	29.79±2.47	97.18±1.86
April	0.70±1.68	23.0±2.00	119.77±0.08	84.21±6.32	129.79±0.36
May	0.79±0.00	26.0±0.06	123.16±0.24	184.61±8.55	187.67±3.25
June	0.84±1.25	27.0±2.14	121.43±0.00	226.43±32.52	225.76±1.97
July	0.98±1.62	24.0±1.18	125.46±0.16	279.86±120.00	534.71±2.40
August	1.07±1.66	22.5±1.62	125.13±0.60	301.76±32.55	486.79±2015
September	1.16±0.82	19.0±1.92	121.31±0.00	331.76±13.72	498.21±52.00
October	1.02±0.15	18.0±1.67	129.46±0.00	189.93±5.88	354.74±11.60
November	0.92±1.25	17.5±1.51	108.76±1.64	101.42±0.80	188.97±62.50
December	0.83±0.00	13.0±0.25	99.79±0.42	64.76±5.62	101.24±0.85
Average ±Sd	0.86±0.17	20.25±4.38	115.41±11.70	154.00±112.91	244.44±177.58

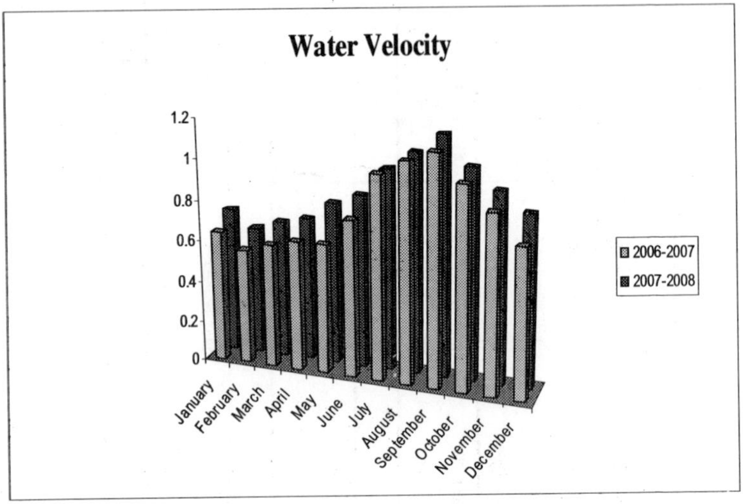

Fig. 5.I.1: Graph showing monthly fluctuation of Water Velocity at Sampling Site-I in 2007-2008

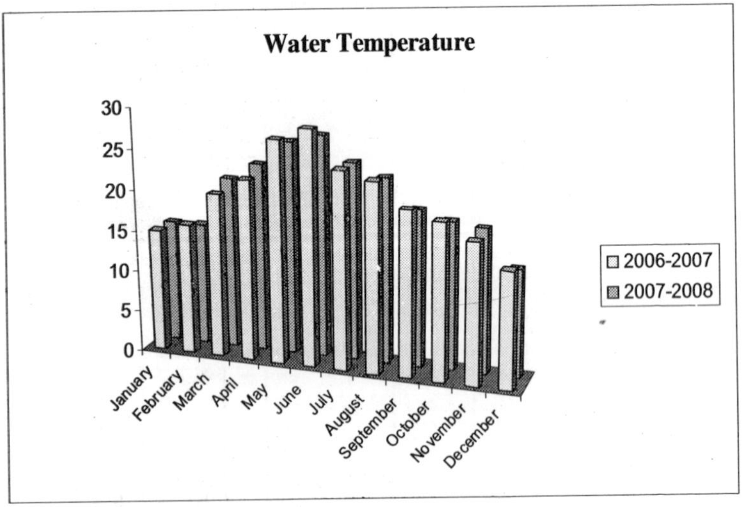

Fig. 5.I.2: Graph showing monthly fluctuation of Water temperature at Sampling Site-I in 2007-2008

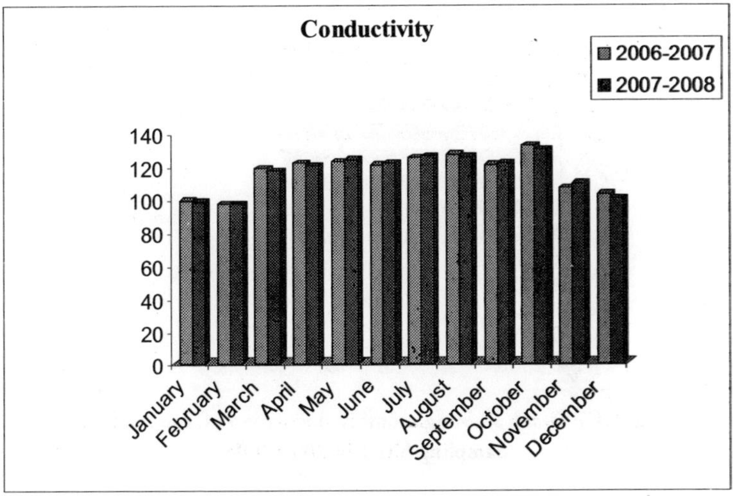

Fig. 5.I.3: Graph showing monthly fluctuation of Conductivity at Sampling Site-I in 2007-2008

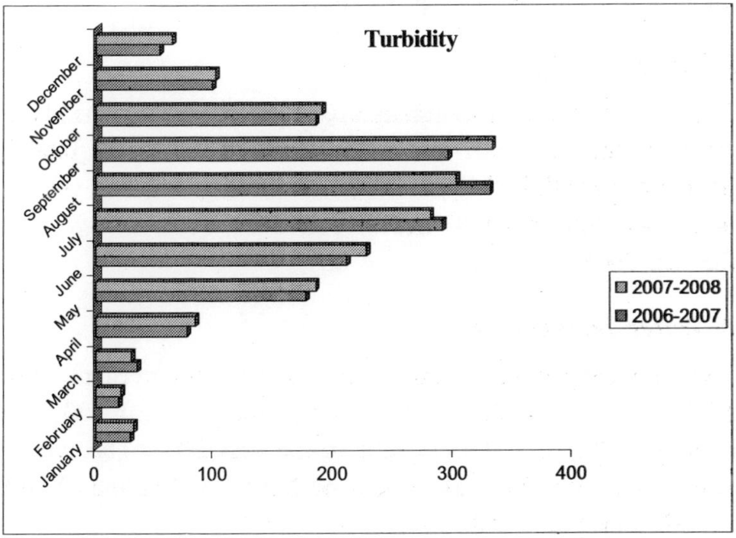

Fig. 5.I.4: Graph showing monthly fluctuation of Turbidity at Sampling Site-I in 2007-2008

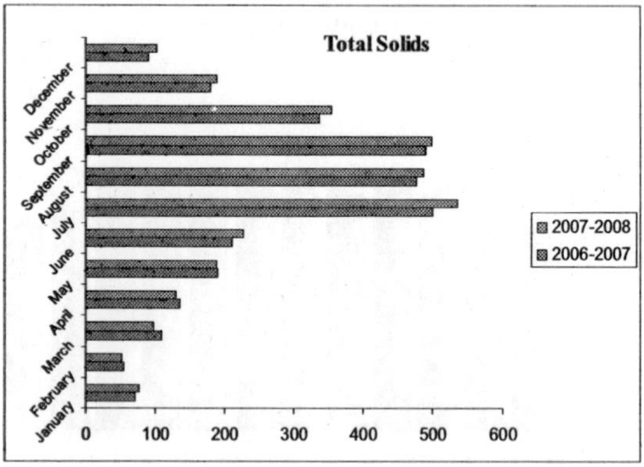

Fig. 5.I.5: Graph showing monthly fluctuation of Total Solid at Sampling Site-I in 2007-2008

Sampling site II

The physical characteristics of sampling site-II are presented in Table 5. I. 2 and Figures Fig. 5. I. 6 to Fig. 5. I. 10.

i. Water Velocity

The velocity of water was minimum 0.58 m/sec ± 0.00 in the March month in the year 2007 and maximum of 1.20 m/sec ± 0.04 in September month in the year 2007. The average value of velocity in the year 2007 was 0.83 m/s ±± 0.20 and 0.88 m/sec. ± 0.19 in the year 2008.

ii. Water Temperature

The water temperature at this sampling site was observed minimum 13.00°C ± 0.26 in the month of December for year 2008 and maximum 28.5°C ± 0.05 in June in the year 2008. The average value in the year 2007 was found to be 20.66°C ± 4.12 and 20.50°C ± 5.12 in the year 2008.

iii. Conductivity

The conductivity of river water was observed to be minimum 96.78 Siemens/cm ± 0.02 in the January month in the year 2008 and maximum of 127.78 Siemens/cm ± 0.35 in the July Month in the year 2008. The average values recorded in the year 2007 was 114.92 Siemens/cm ± 9.51 and 115.15 Siemens/cm ± 11.40 in the year 2008.

iv. Turbidity

The turbidity in water was minimum 16.79 JTU ± 1.91 in the month of February in the year 2007 and maximum of 376.46 JTU ± 178.06 in the month of August in the year 2007. The average value of turbidity in the year 2007 was 159.98 JTU ± 123.19 and 165.79 JTU ± 120.65 in the year 2008.

v. Total Solids

The total solids were minimum 52.93 mg/l ± 0.21 in February month in the year 2007 and maximum of 529.17 mg/l ± 0.36 in the September month in the year 2008. The average value of total solids in the year 2007 was 234.34 mg/l ± 150.53 and 247.65 mg/l ± 167.36 in the year 2008.

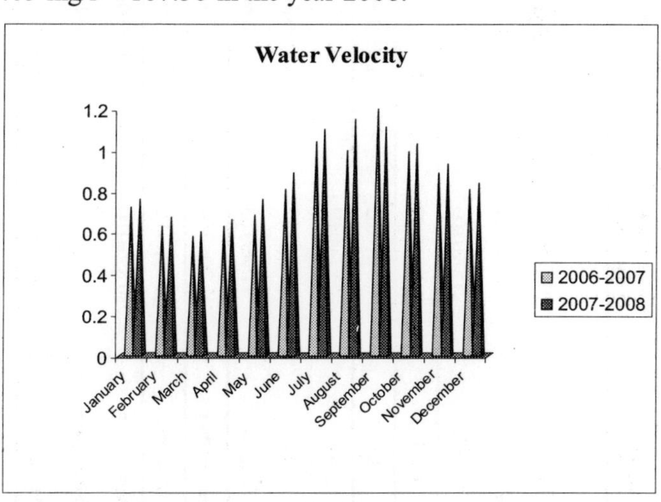

Fig. 5.I.6: Graph showing monthly fluctuation of Water Velocity at Sampling Site-II in 2007-2008

Table 5. II. 2: Monthly fluctuation in Physical parameters of Sampling Site – II during 2007-2008

Months	Water velocity (m/sec)	Water temperature (°C)	Conductivity (Siemens/cm)	Turbidity (JTU)	Total solids (mg/l)
2007 (Sampling Station – II)					
January	0.72±0.01	16.00±0.00	101.59±0.03	25.36±2.06	71.51±0.32
February	0.63±0.02	18.00±0.25	99.54±0.10	16.79±1.91	52.93±0.21
March	0.58±0.00	20.50±1.25	106.75±0.02	28.94±7.46	112.93±0.04
April	0.63±0.05	21.00±1.05	116.75±0.00	66.79±6.81	164.90±1.46
May	0.68±0.02	26.00±0.50	126.46±0.00	180.46±62.00	187.14±0.58
June	0.81±0.02	27.40±0.37	123.41±0.01	215.70±71.51	206.47±0.18
July	1.04±0.03	25.00±0.23	119.50±0.15	296.49±77.41	350.75±0.24
August	1.00±0.03	23.50±0.60	122.78±0.03	376.46±178.06	442.64±0.06
September	1.20±0.04	21.00±0.40	118.79±0.01	302.95±3.00	510.15±0.08
October	0.99±0.01	18.00±1.06	125.76±0.03	215.46±27.00	369.71±0.20
November	0.89±0.00	16.50±0.00	111.78±0.03	128.95±2.80	234.78±0.80
December	0.81±0.06	15.00±0.15	105.95±0.05	65.46±7.35	108.15±0.21
Average ±Sd	0.83±0.20	20.66±4.12	114.92±9.51	159.98±123.19	234.34±150.53

contd. on next page...

contd...

2008 (Sampling Station – II)					
January	0.76±0.12	14.00±0.32	96.78±0.02	39.15±1.25	82.25±0.25
February	0.67±0.02	16.00±0.37	99.92±0.01	21.79±0.90	63.10±25.50
March	0.60±0.00	19.00±0.05	104.95±0.12	30.25±0.05	105.32±2.35
April	0.66±0.05	22.50±0.35	114.35±0.10	56.12±2.15	159.36±0.25
May	0.76±0.00	27.00±0.91	126.78±0.00	192.79±6.34	190.18±11.50
June	0.89±0.20	28.50±0.05	124.67±0.03	235.49±14.64	236.61±3.52
July	1.10±0.00	26.00±3.90	127.78±0.35	285.71±18.27	411.15±32.51
August	1.15±0.25	24.00±2.12	123.98±0.52	365.95±30.25	498.70±0.32
September	1.11±0.03	20.00±1.50	120.10±0.10	309.12±18.35	529.17±0.36
October	1.03±0.02	19.00±0.25	125.95±0.00	235.17±2.65	398.72±1.35
November	0.93±0.02	17.00±0.17	113.54±0.04	132.48±5.00	192.12±10.30
December	0.84±0.14	13.00±0.26	102.98±0.03	85.48±4.32	105.15±0.15
Average ±Sd	0.88±0.19	20.50±5.12	115.15±11.40	165.79±120.65	247.65±167.36

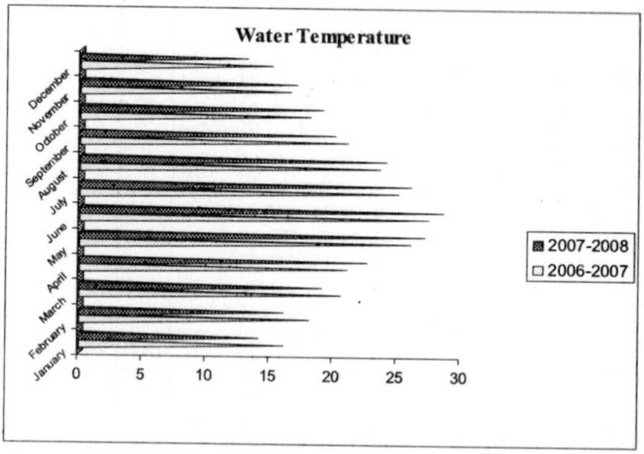

Fig. 5.II.7: Graph showing monthly fluctuation of Water temperature at Sampling Site-II in 2007-2008

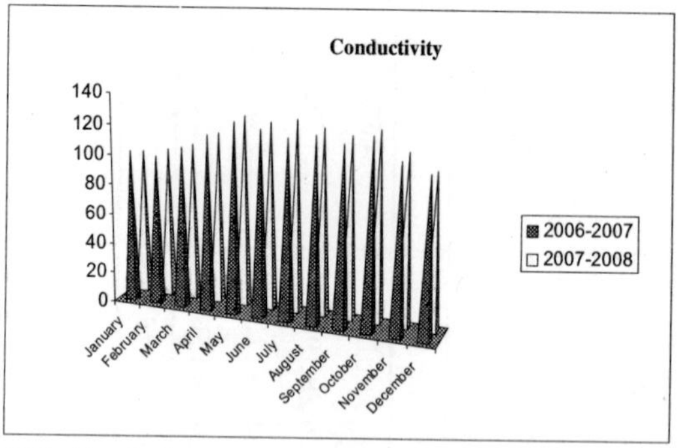

Fig. 5.II.8: Graph showing monthly fluctuation of Conductivity at Sampling Site-II in 2007-2008

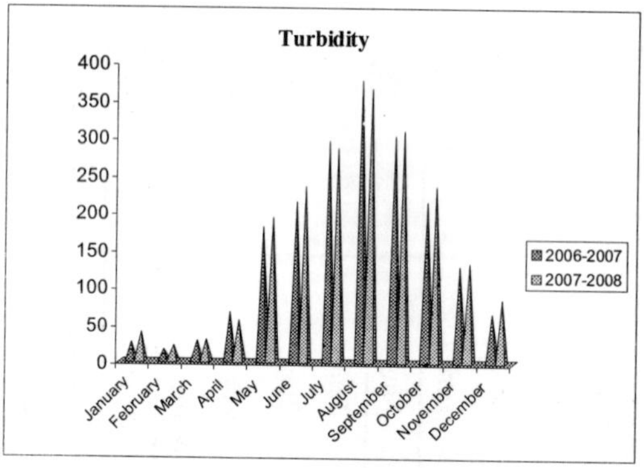

Fig. 5.II.9: Graph showing monthly fluctuation of Turbidity at Sampling Site-II in 2007-2008

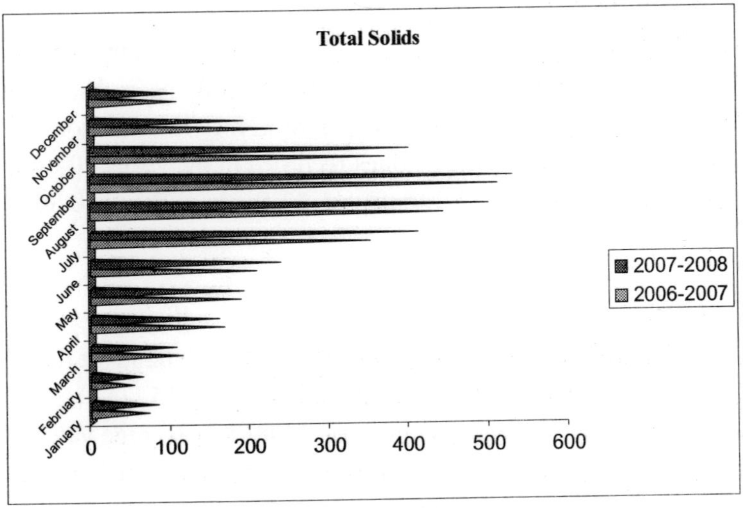

Fig. 5.II.10: Graph showing monthly fluctuation of Total Solid at Sampling Site-II in 2007-2008

Sampling site III

The physical characteristics of sampling site- III are presented in Table 5. III. 3 and Figs. 5. III. 11 to Figs. 5.III. 15

i. Water Velocity

The velocity of water was minimum 0.61 m/sec ± 0.31 in the February month in the year 2007 and maximum of 1.13 m/sec ± 0.18 in September month in the year 2007. The average value of velocity in the year 2007 was 0.84 m/s ± 0.19 and 0.88 m/sec. ± 0.16 in the year 2008.

ii. Water Temperature

The water temperature at this sampling site was observed minimum 13.00°C ± 0.35 in the month of December for year 2007 and maximum 29.00°C ± 0.50 in June in the year 2007. The average value in the year 2007 was found to be 20.33°C ± 5.08 and 20.83°C ± 4.23 in the year 2008.

iii. Conductivity

The conductivity of river water was observed to be minimum 96.45 Siemens/cm ± 0.05 in the February month in the year 2007 and maximum of 128.47 Siemens/cm ± 0.00 in the October month in the year 2008. The average values recorded in the year 2007 was 116.04 Siemens/cm ± 22.29 and 116.25 Siemens/cm ± 10.30 in the year 2008.

iv. Turbidity

The turbidity in water was minimum 21.84 JTU ± 0.02 in the month of February in the year 2008 and maximum of 349.15 JTU ± 25.00 in the month of September in the year 2008. The average value of turbidity in the year 2007 was 159.01 JTU ± 104.17 and 172.13 JTU ± 116.99 in the year 2008.

v. Total Solids

The total solids were minimum 59.71 mg/l ± 0.35 in February month in the year 2007 and maximum of 505.17 mg/l ± 0.18 in the September month in the year 2008. The average value of total solids in the year 2007 was 245.55 mg/l ± 163.41 and 249.78 mg/l ± 170.56 in the year 2008.

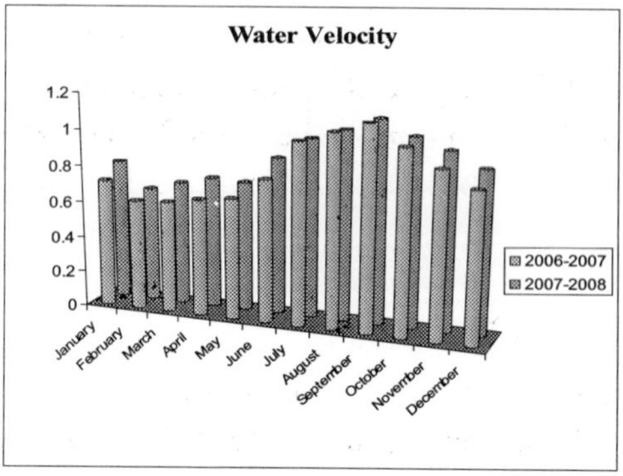

Fig. 5.III.11: Graph showing monthly fluctuation of Water Velocity at Sampling Site-III in 2007-2008

Table 5. III. 3: Monthly fluctuation in Physical parameters of Sampling Site - III during 2007-2008

Months	Water velocity (m/sec)	Water temperature (°C)	Conductivity (Siemens/cm)	Turbidity (JTU)	Total solids (mg/l)
2007 (Sampling Station - III)					
January	0.71±1.15	14.0±0.01	97.69±0.01	42.98±0.86	79.84±0.18
February	0.61±0.31	16.0±0.06	96.45±0.05	22.17±0.00	59.71±0.35
March	0.62±0.30	20.0±1.07	109.19±0.02	42.96±1.00	102.51±0.19
April	0.65±0.19	23.0±1.05	116.75±0.00	82.72±7.81	138.73±0.28
May	0.67±0.00	28.0±0.07	124.28±0.01	180.17±7.86	184.21±0.26
June	0.79±1.13	29.0±0.50	122.64±0.00	205.48±37.00	244.41±0.08
July	1.01±0.75	23.0±0.27	125.46±0.24	279.71±127.86	484.69±0.19
August	1.07±0.06	22.0±1.89	127.92±0.01	325.14±61.28	451.17±1.98
September	1.13±0.18	20.5±0.50	123.75±0.03	289.13±7.84	498.67±46.00
October	1.02±0.73	19.4±0.03	128.16±0.09	203.45±22.07	369.23±12.05
November	0.92±1.18	16.0±0.00	112.91±0.05	135.75±7.09	215.75±72.50
December	0.83±0.20	13.0±0.35	107.27±0.02	98.45±4.86	117.64±0.14
Average ±Sd	0.84±0.19	20.33±5.08	116.04±22.29	159.01±104.17	245.55±163.41

contd. on next page...

contd...

	2008 (Sampling Station - III)				
January	0.78±0.01	15.0±0.04	99.73±0.00	59.41±2.10	81.71±0.06
February	0.64±0.03	17.5±0.15	100.18±0.16	21.84±0.02	62.79±0.09
March	0.69±0.05	22.5±0.56	110.37±0.05	31.84±0.05	92.17±0.03
April	0.73±0.00	24.0±1.06	115.17±0.04	87.28±6.12	124.41±2.15
May	0.72±0.02	27.0±0.26	123.78±0.03	182.31±3.16	186.41±0.16
June	0.87±0.22	26.0±0.20	123.10±0.21	245.75±52.05	239.79±0.16
July	0.99±0.03	24.0±0.15	124.98±0.11	286.75±158.25	498.95±0.32
August	1.05±0.05	22.5±0.60	126.21±0.21	332.45±22.46	479.49±0.12
September	1.12±0.12	21.0±0.50	122.49±0.04	349.15±25.00	505.17±0.18
October	1.04±2.15	19.0±1.21	128.47±0.00	242.10±2.10	375.24±0.50
November	0.98±0.04	17.5±0.05	115.75±0.90	134.42±2.52	227.10±0.15
December	0.90±0.06	14.0±0.10	104.71±0.05	92.31±5.12	124.18±2.15
Average ±Sd	0.88±0.16	20.83±4.23	116.25±10.30	172.13±116.99	249.78±170.56

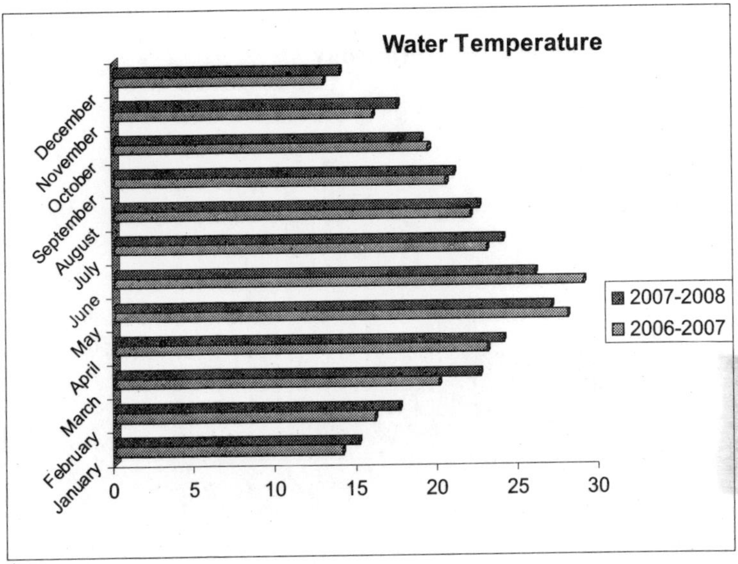

Fig. 5.III.12: Graph showing monthly fluctuation of Water temperature at Sampling Site-III in 2007-2008

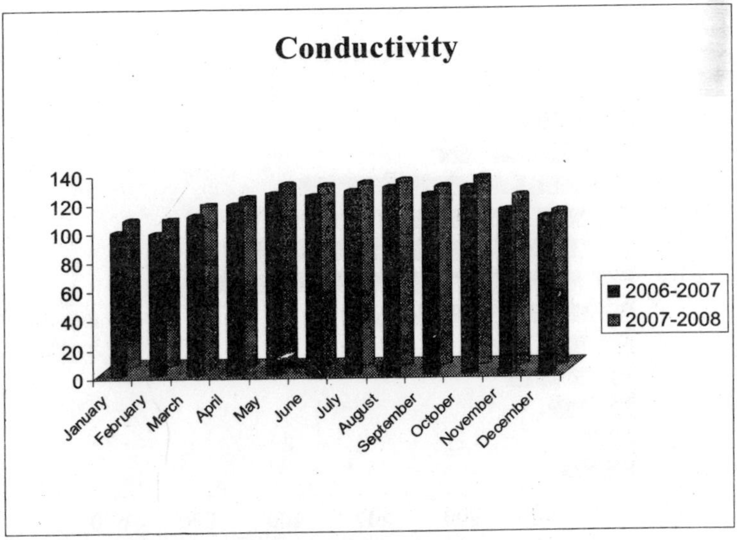

Fig. 5.III.13: Graph showing monthly fluctuation of Conductivity at Sampling Site-III in 2007-2008

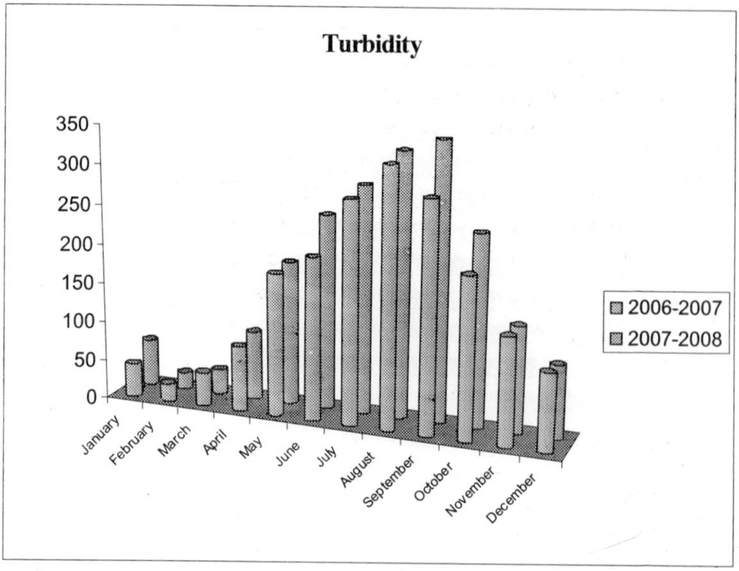

Fig. 5.III.14: Graph showing monthly fluctuation of Turbidity at Sampling Site-III in 2007-2008

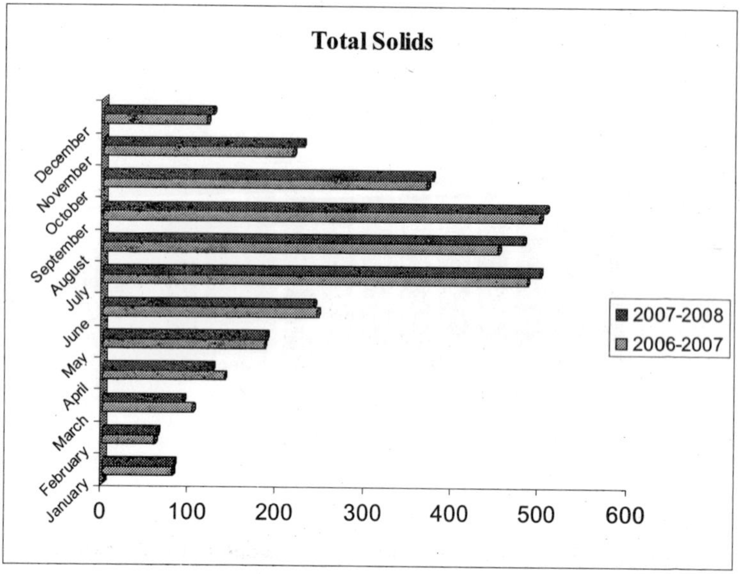

Fig. 5.III.15: Graph showing monthly fluctuation of Total Solid at Sampling Site-III in 2007-2008

CHAPTER-6

CHEMICAL ANALYSIS

The chemical analysis of water provides considerable insight into the health and workings of lakes, rivers, oceans, and groundwater. Water chemistry has helped scientists to define the different currents and circulation of the world's oceans, improved their understanding of water's interactions with Earth's geologic materials, and given insight into the impact of human activities on water bodies. It also has provided a clearer understanding of the limits of a water body's ability to assimilate (take in) some level of pollution without harming the water system, its aquatic plants and animals, and humans who may use the water. During the current study some chemical parameters were studied which are mentioned (Sampling site-wise) below:-

Sampling site I

The chemical characteristics of sampling site- I are presented in Table 6.I. 1 and Figs. 6.I.1 to 6.I.9.

i. Dissolved Oxygen (DO)

Dissolved oxygen (DO) in water was found minimum 7.1 mg/l ± 0.16 in the April month in 2008 and maximum 10.4 mg/l ± 0.17 in the December month in the year 2008. The average value of DO in the year 2007 was observed to be 8.83 mg/l ± 0.86 and 8.78 mg/l ± 0.92 in the year 2008.

ii. Free CO_2

Free CO_2 of river water was found to be minimum with 0.1 mg/l ± 0.15 in the month of January in the year 2007 and maximum of 1.4 mg/l ± 1.4 in the month of April in the year 2008 .The average value of free CO_2 in the year 2007 was observed 0.8 mg/l ± 0.75 and 0.7 mg/l ± 0.73 in the year 2008.

iii. Acidity

The acidity of water was observed minimum of 27.54 mg/l ± 0.06 in November month in the year 2008 and maximum of 60.32 mg/l± 0.00 in the month of February in the year 2007. The average value observed in the year 2007 was 45.47 mg/l ± 10.22 and 44.18 mg/l ± 9.11 in the year 2008.

iv. Total Alkalinity

The total alkalinity of river water at this sampling station was minimum 109.00 mg/l ± 0.12 in the month of September in the year 2007 and maximum of 220.00 mg/l ± 0.09 in the month of May in the year 2007. The average value in the year 2007 was 173.67 mg/l ± 36.63 and 176.25 mg/l ± 26.34 in the year 2008.

v. Total Hardness

The total hardness of water was recorded minimum 112.50 mg/l ± 0.00 in the month of March in the year 2007 and maximum value of 245.00 mg/l ± 8.10 was recorded in the month of May in the year 2008. The average value observed in the year 2007 was 170.46 mg/l ± 33.51 and 167.59 mg/l ± 37.27 in the year 2008.

vi. Calcium Hardness

The calcium hardness in river water at sampling site - I was minimum 19.1 mg/l ± 0.30 in month of March in the year 2008 and maximum of 46.7 mg/l ± 2.30 in February in the year 2007. The average value in the year 2007 was 31.59 mg/l ± 8.03 and 31.63 mg/l ± 8.28 in the year 2008.

vii. Magnesium Hardness

The magnesium hardness in river water at sampling site-I was found minimum 11.89 mg/l ± 0.24 in the month of March in the year 2007 and maximum of 27.30 mg/l ± 3.31 in the month of June in the year 2008. The average value in the year 2007 was observed 19.20 mg/l ± 4.53 and 19.58 mg/l ± 4.43 in the year 2008.

viii. Chlorides

The Chlorides in water was minimum 5.01 mg/l± 0.21 in the month of July in the year 2007 and maximum of 13.21 mg/l ± 3.51 in the month of October in the year 2008. The average value of chlorides in the year 2007 was 7.92 mg/l ± 1.87 and 8.22 mg/l ± 1.81 in the year 2008.

ix. Phosphates

At this site the phosphate was recorded to be minimum with 0.26 mg/l ± of October month in year 2007 and was maximum 0.84 mg/l ± of May month in year 2008. The annual mean calculated for year 2007 was 0.52 mg/l ± 0.20 and of year 2008 was 0.51 mg/l ± 0.18.

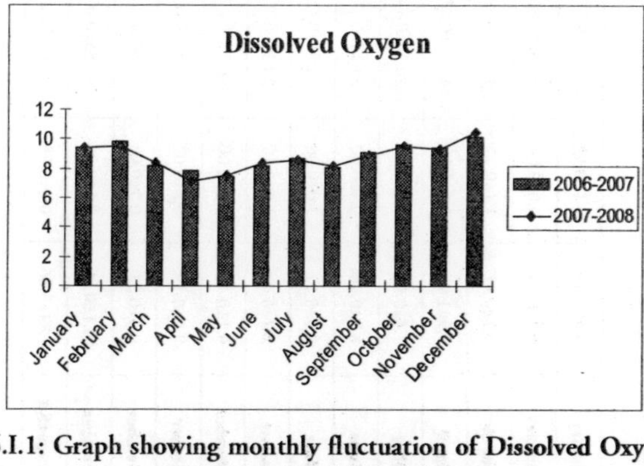

Fig. 6.I.1: Graph showing monthly fluctuation of Dissolved Oxygen at Sampling Site-I in 2007-2008

Table 6.I.1: Monthly fluctuation in Chemical parameters of Sampling Site – I during 2007-2008

Months	DO (mg/l)	Free CO_2 (mg/l)	Acidity (mg/l)	Total Alkalinity (mg/l)	Total Hardness (mg/l)	Ca Hardness (mg/l)	Mg Hardness (mg/l)	Chlorides (mg/l)	Phosphates (mg/l)
2007 (Sampling Station – I)									
January	9.4±0.00	0.1±0.15	57.22±0.32	195.0±3.57	158.0±1.19	34.80±0.03	19.50±5.64	8.80±0.16	0.69±0.17
February	9.8±0.03	0.6±0.04	60.32±0.00	201.5±0.05	133.0±0.23	46.70±2.30	14.90±10.07	8.28±0.50	0.54±0.19
March	8.2±0.82	1.1±0.12	44.21±0.14	135.0±4.75	112.5±0.00	21.20±0.05	11.89±0.24	9.06±0.09	0.29±0.15
April	7.9±1.03	1.6±1.18	47.73±0.31	148.5±1.03	142.5±0.05	23.50±0.00	13.76±0.28	7.59±0.03	0.56±0.21
May	7.4±0.05	0.9±0.10	50.22±0.14	220.0±0.09	218.0±0.11	27.50±0.22	22.70±0.54	6.10±0.10	0.82±0.20
June	8.2±0.06	0.4±0.00	54.12±0.17	210.0±0.12	214.0±0.59	24.28±0.22	26.40±1.80	6.76±0.34	0.79±0.06
July	8.7±0.12	1.0±0.00	42.51±0.05	164.5±6.91	175.0±0.12	31.54±0.34	19.10±1.07	5.01±0.21	0.37±0.05
August	8.1±0.40	1.0±0.10	28.25±0.17	125.0±0.17	189.0±0.48	38.50±0.22	21.00±1.27	7.82±0.65	0.43±0.10
September	9.1±0.17	0.8±0.00	46.71±0.12	109.0±0.12	148.5±0.23	29.70±0.15	18.90±0.39	7.78±0.20	0.32±0.16
October	9.6±0.19	0.8±0.02	34.52±0.00	189.0±4.75	163.0±0.00	34.50±0.06	19.20±0.29	12.67±0.05	0.26±0.09
November	9.4±0.23	0.9±0.00	30.00±1.79	181.0±7.39	208.0±0.17	42.40±1.15	16.70±4.77	7.90±0.00	0.41±0.10
December	10.1±0.08	0.6±0.00	49.85±0.03	205.5±0.10	184.0±0.06	24.50±0.08	26.30±0.10	7.29±0.05	0.76±0.12
Average±Sd	8.83±0.86	0.8±0.75	45.47±10.22	173.67±36.63	170.46±33.51	31.59±8.03	19.20±4.53	7.92±1.87	0.52±0.20

contd. on next page...

contd...

2008 (Sampling Station – I)									
January	9.4±0.81	0.2±0.00	52.76±1.24	183.0±1.65	148.0±1.65	33.10±2.14	18.40±2.38	8.42±1.06	0.58±0.18
February	9.5±0.80	0.2±0.12	58.19±1.65	203.0±1.68	132.0±0.16	44.90±1.65	15.10±7.25	7.92±2.48	0.49±0.21
March	8.4±1.56	0.9±0.05	41.49±2.54	163.0±0.24	123.5±1.72	19.10±0.30	12.64±6.91	8.98±1.68	0.32±0.25
April	7.1±0.16	1.4±1.14	46.00±1.68	172.0±2.16	137.0±6.12	22.90±0.06	14.62±29.48	7.79±1.77	0.63±0.22
May	7.6±0.02	1.1±1.64	49.46±0.11	208.0±0.14	245.0±8.10	26.30±2.14	24.50±29.00	6.45±1.87	0.84±0.20
June	8.4±1.64	0.3±0.00	55.27±0.25	196.0±0.16	212.0±1.00	22.10±1.68	27.30±3.31	6.98±1.62	0.76±0.21
July	8.6±0.00	0.7±0.00	44.91±1.68	172.0±0.04	164.0±2.80	32.50±1.62	19.90±5.74	6.12±0.25	0.33±0.15
August	8.2±0.20	0.8±0.00	37.48±1.25	143.0±1.68	189.2±2.15	39.40±1.06	22.10±5.30	7.83±1.54	0.48±0.16
September	8.9±1.16	0.1±0.10	44.91±0.11	121.0±0.14	132.4±1.64	31.50±1.60	19.00±0.19	8.23±1.62	0.35±0.09
October	9.5±0.15	0.6±0.00	32.35±1.60	179.0±5.64	157.0±0.01	35.60±1.25	19.90±6.03	13.21±3.51	0.28±0.10
November	9.3±2.14	0.2±0.10	27.54±0.06	167.0±1.68	198.0±0.16	40.50±0.12	17.00±0.19	9.13±1.62	0.43±0.12
December	10.4±0.17	0.4±0.00	39.79±0.13	208.0±1.40	173.0±2.46	27.50±2.46	24.50±0.00	7.63±0.14	0.63±0.19
Average ±Sd	8.78±0.92	0.7±0.73	44.18±9.11	176.25±26.34	167.59±37.27	31.63±8.28	19.58±4.43	8.22±1.81	0.51±0.18

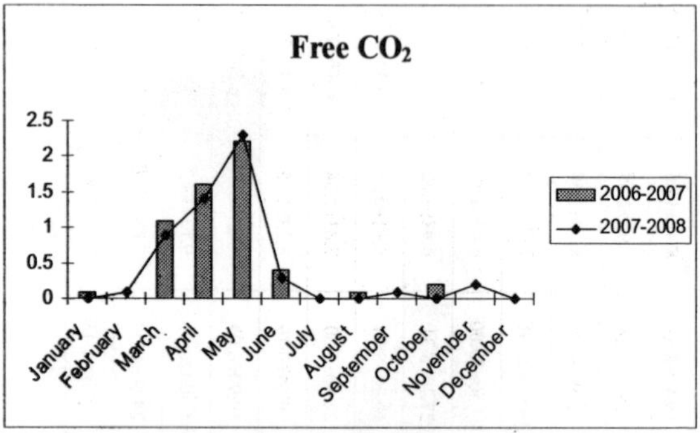

Fig. 6.I.2: Graph showing monthly fluctuation of Free CO_2 at Sampling Site-I in 2007-2008

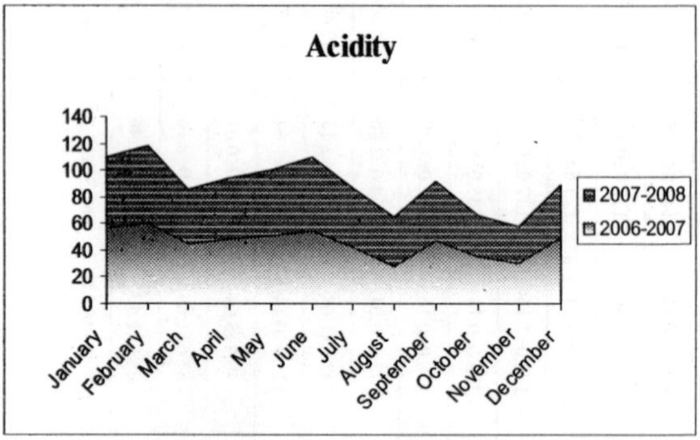

Fig. 6.I.3: Graph showing monthly fluctuation of Acidity at Sampling Site-I in 2007-2008

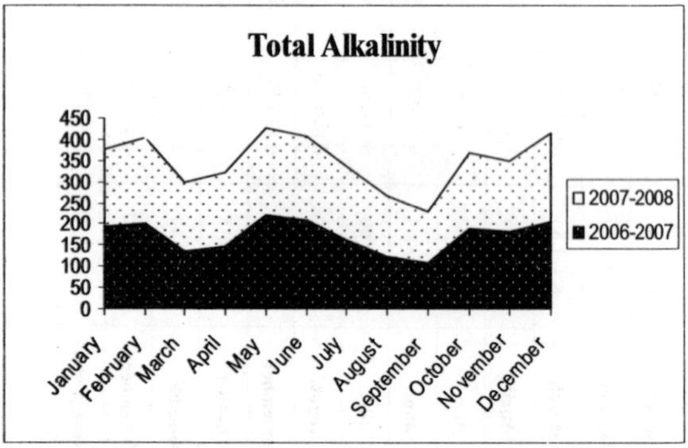

Fig. 6.I.4: Graph showing monthly fluctuation of Total Alkalinity at Sampling Site-I in 2007-2008

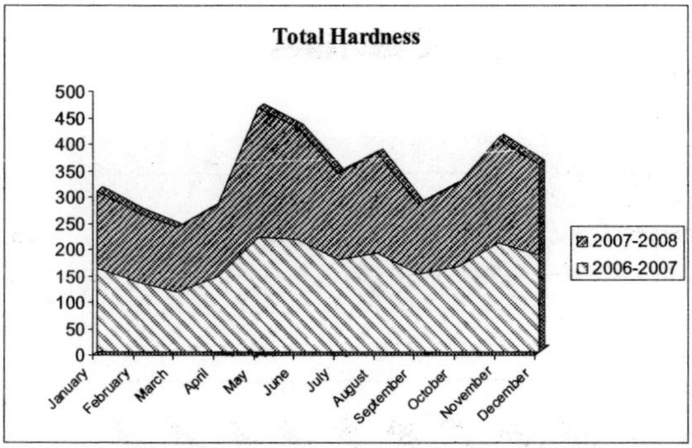

Fig. 6.I.5: Graph showing monthly fluctuation of Total Hardness at Sampling Site-I in 2007-2008

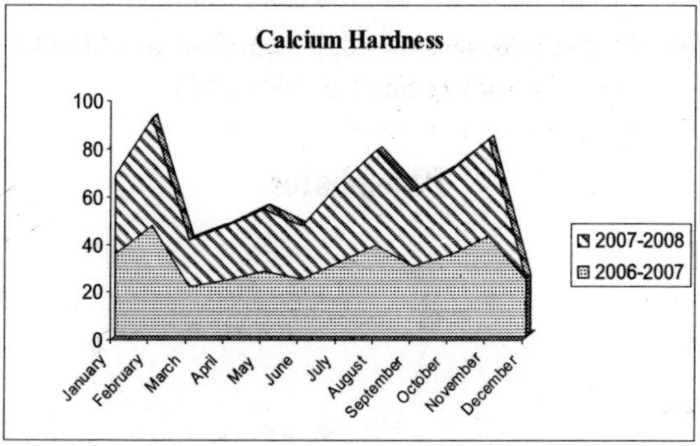

Fig. 6.I.6: Graph showing monthly fluctuation of Calcium Hardness at Sampling Site-I in 2007-2008

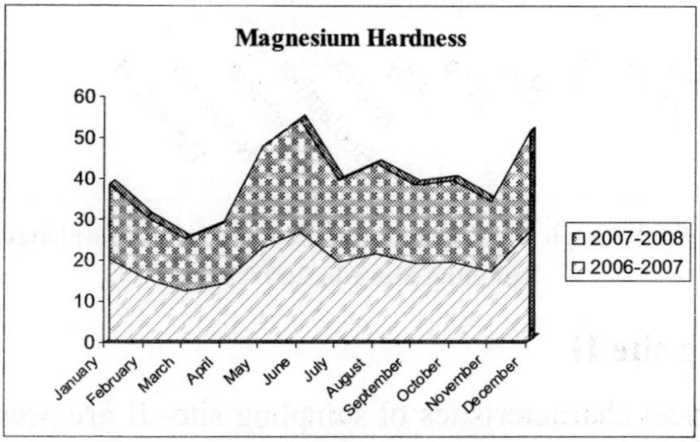

Fig. 6.I.7: Graph showing monthly fluctuation of Magnesium Hardness at Sampling Site-I in 2007-2008

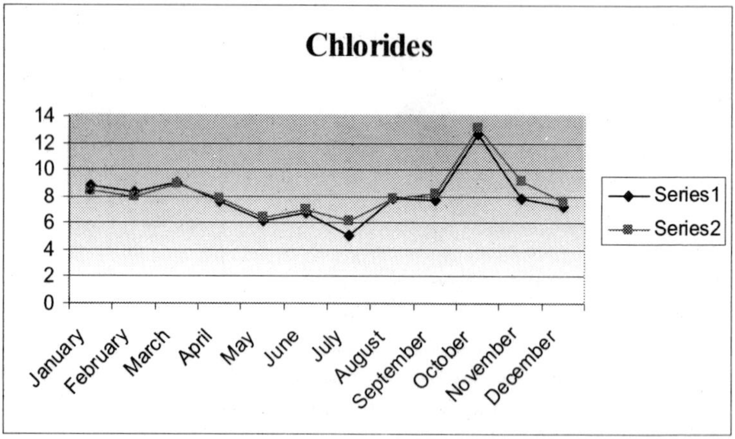

Fig. 6.I.8: Graph showing monthly fluctuation of Chlorides at Sampling Site-I in 2007-2008

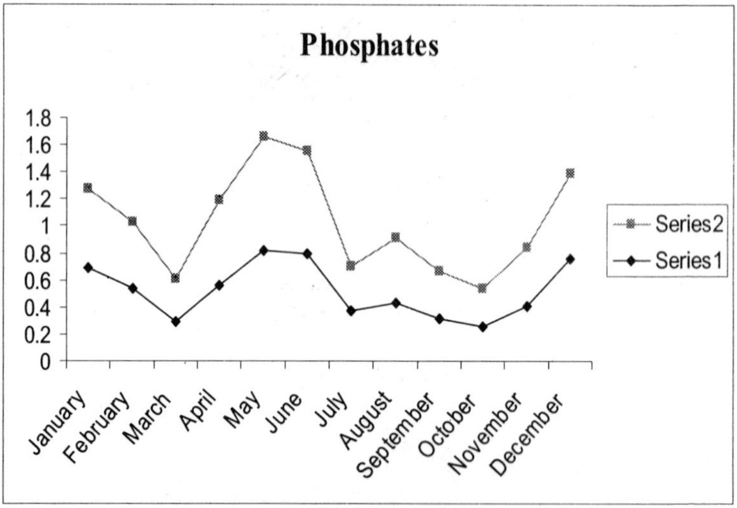

Fig. 6.I.9: Graph showing monthly fluctuation of Phosphates at Sampling Site-I in 2007-2008

Sampling site II

The chemical characteristics of sampling site- II are presented in Table 6. II. 2 and Figs. 6.II.10 to 6.II.18.

i. Dissolved Oxygen (DO)

Dissolved oxygen (DO) in water was found minimum 7.5 mg/l ± 0.05 in the May month in 2008 and maximum of 10.5 mg/l ± 0.05 in the January month in the year 2008. The average value of DO in the year 2007 was observed to be 8.98 mg/l ± 0.87 and 9.04 mg/l ± 0.90 in the year 2008.

ii. Free CO_2

Free CO_2 of river water was found to be minimum with 0.2 mg/l ± 0.00 in the month of January in the year 2007 and maximum of 1.9 mg/l ± 0.05 in the month of May in the year 2007. The average value of free CO_2 in the year 2007 was observed 0.6 mg/l ± 0.73 and 0.7 mg/l ± 0.72 in the year 2008.

iii. Acidity

The acidity of water was observed minimum of 26.51 mg/l ± 0.26 in December month in the year 2008 and maximum of 58.62 mg/l± 0.03 in the month of February in the year 2007. The average value observed in the year 2007 was 43.81 mg/l ± 8.29 and 42.33 mg/l ± 8.82 in the year 2008.

iv. Total Alkalinity

The total alkalinity of river water at this sampling station was minimum 115.00 mg/l ± 0.85 in the month of September in the year 2007 and maximum of 213.00 mg/l ± 10.25 in the month of May in the year 2008 .The average value in the year 2007 was 175.31 mg/l ± 30.84 and 178.92 mg/l ± 27.83 in the year 2008.

v. Total Hardness

The total hardness of water was recorded minimum 117.00 mg/l ± 0.00 in the month of March in the year 2007 and maximum value of 226.00 mg/l ± 3.56 was recorded in the month of May in the year 2008. The average value observed in the year 2007 was 169.83 mg/l ± 30.37 and 170.50 mg/l ± 29.75 in the year 2008.

vi. Calcium Hardness

The calcium hardness in river water at sampling site-II was minimum 21.91 mg/l ± 6.10 in month of June in the year 2008 and maximum of 44.95 mg/l ± 0.09 in February in the year 2007. The average value in the year 2007 was 33.45 mg/l ± 6.25 and 31.42 mg/l ± 7.02 in the year 2008.

vii. Magnesium Hardness

The magnesium hardness in river water at sampling site- II was found minimum 12.39 mg/l ± 0.15 in the month of March in the year 2007 and maximum of 27.82 mg/l ± 0.59 in the month of June in the year 2007. The average value in the year 2007 was observed 19.91 mg/l ± 4.52 and 20.88 mg/l ± 3.95 in the year 2008.

viii. Chlorides

The chlorides in water was minimum 5.65 mg/l± 0.21 in the month of July in the year 2007 and maximum of 12.45 mg/l ± 0.24 in the month of October in the year 2008. The average value of Chlorides in the year 2007 was 7.88 mg/l ± 1.29 and 7.94 mg/l ± 1.72 in the year 2008.

ix. Phosphates

At this site the phosphate was recorded to be minimum with 0.28 mg/l ± in October month in year 2008 and was maximum 0.92 mg/l ± of May month in year 2007. The annual mean calculated for year 2007 was 0.53 mg/l ± 0.20 and of year 2008 was 0.55 mg/l ± 0.19.

Table 6.II.2: Monthly fluctuation in Chemical parameters of Sampling Site – II during 2007-2008

Months	DO (mg/l)	Free CO_2 (mg/l)	Acidity (mg/l)	Total Alkalinity (mg/l)	Total Hardness (mg/l)	Ca Hardness (mg/l)	Mg Hardness (mg/l)	Chlorides (mg/l)	Phosphates (mg/l)
2007 (Sampling Station – II)									
January	10.3±0.00	0.3±0.03	53.99±0.05	194.0±2.25	161.0±0.02	37.45±0.07	19.85±0.24	8.64±0.03	0.55 ± 0.10
February	9.5±0.03	0.6±0.00	58.62±0.03	199.5±3.55	139.0±0.09	44.95±0.09	15.26±0.39	8.10±0.00	0.47 ±0.15
March	8.8±0.02	1.3±0.06	46.20±0.06	159.4±0.06	117.0±0.00	28.78±0.03	12.39±0.15	8.98±0.02	0.38 ±0.17
April	7.9±0.03	1.7±0.03	44.69±0.02	147.0±0.09	138.0±0.05	24.60±0.00	14.85±0.24	7.92±0.06	0.61 ±0.14
May	7.6±0.17	1.9±0.05	45.52±0.17	212.0±0.00	216.0±0.01	26.40±0.21	23.46±0.58	6.56±0.05	0.92 ±0.19
June	8.3±0.24	0.3±0.27	49.32±0.07	205.0±0.12	213.0±0.40	24.96±0.52	27.82±0.59	6.89±0.30	0.86 ±0.18
July	8.8±0.07	0.6±0.00	43.21±0.10	179.2±2.23	183.0±0.09	29.15±0.50	21.42±1.99	5.65±0.21	0.37 ±0.11
August	8.3±0.09	0.4±0.49	31.94±0.08	137.0±0.01	191.0±0.71	35.14±0.12	23.00±2.87	6.99±0.75	0.46 ±0.06
September	8.9±0.26	0.7±0.00	44.11±0.00	115.0±0.85	152.0±0.23	31.78±0.88	19.87±3.16	7.76±0.20	0.37 ±0.07
October	9.4±0.53	0.5±0.00	37.56±0.05	164.0±0.27	167.0±0.02	34.23±0.06	18.95±0.19	10.58±0.03	0.28 ±0.12
November	9.7±0.07	0.5±0.00	29.52±0.02	187.0±1.47	189.0±0.07	40.12±0.04	17.12±0.15	8.67±0.07	0.40 ±0.14
December	10.2±0.03	0.4±0.00	41.00±0.19	204.6±0.00	172.0±0.10	31.78±0.05	24.96±0.24	7.86±0.11	0.64 ±0.15
Average±Sd	8.98±0.87	0.6±0.73	43.81±8.29	175.31±30.84	169.83±30.37	33.45±6.25	19.91±4.52	7.88±1.29	0.53±0.20

contd. on next page...

contd...

2008 (Sampling Station – II)								
January	10.5±0.05	0.2±0.00	44.12±0.24	204.0±5.64	153.0±10.35	35.92±0.21	21.75±0.85	8.96±7.10	0.54 ± 0.09
February	9.7±0.08	0.5±0.01	54.69±0.80	209.0±0.12	147.0±0.25	42.15±0.10	17.95±0.73	7.91±8.07	0.47 ± 0.12
March	8.9±0.20	1.1±0.21	45.97±0.30	179.0±0.68	131.0±8.20	26.45±2.50	13.74±2.24	8.46±2.00	0.33 ± 0.14
April	8.1±0.02	1.5±0.13	46.10±3.02	158.0±2.16	139.0±8.61	21.96±8.01	15.98±0.15	7.35±6.10	0.70 ± 0.15
May	7.5±0.05	1.8±1.02	47.21±0.31	213.0±10.25	226.0±3.56	24.69±0.02	25.64±0.58	6.79±0.15	0.90 ± 0.18
June	8.5±0.25	0.7±0.02	54.31±0.24	201.0±3.25	219.0±0.26	21.91±6.10	27.12±0.22	7.10±6.25	0.86 ± 0.16
July	8.7±0.16	0.5±0.00	45.54±0.25	182.0±0.26	173.0±3.08	27.89±0.24	22.63±0.81	5.89±15.00	0.45 ± 0.20
August	8.4±0.10	0.6±0.00	35.10±5.60	149.0±0.35	191.0±8.26	36.45±3.10	23.54±0.26	6.54±2.50	0.59 ± 0.21
September	8.8±0.35	0.7±0.00	42.12±0.30	123.0±0.51	163.0±12.56	30.12±0.25	20.96±0.48	7.59±6.08	0.42 ± 0.15
October	9.6±0.20	0.4±0.00	36.52±7.06	157.0±30.15	154.0±0.21	36.15±0.16	19.12±0.11	12.45±0.24	0.34 ± 0.11
November	9.5±0.03	0.5±0.00	29.78±0.24	174.0±7.26	182.0±3.01	41.14±3.52	18.62±0.93	9.13±2.60	0.43 ± 0.09
December	10.3±0.32	0.3±0.00	26.51±0.26	198.0±35.01	168.0±0.35	32.17±0.26	23.54±0.51	7.15±0.25	0.61 ± 0.11
Average±Sd	9.04±0.90	0.7±0.72	42.33±8.82	178.92±27.83	170.50±29.75	31.42±7.02	20.88±3.95	7.94±1.72	0.55± 0.19

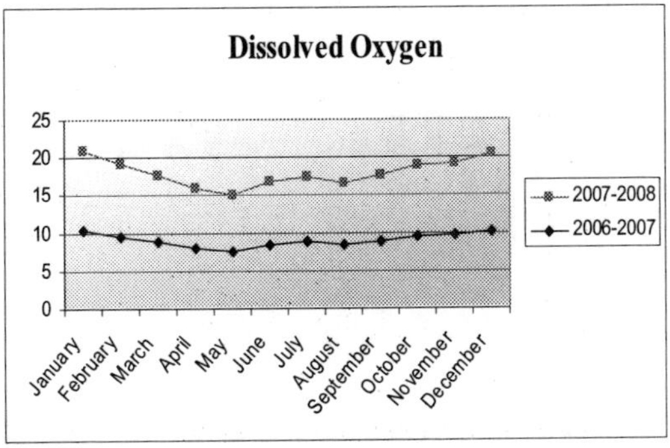

Fig. 6.II.10: Graph showing monthly fluctuation of Dissolved Oxygen at Sampling Site-II in 2007-2008

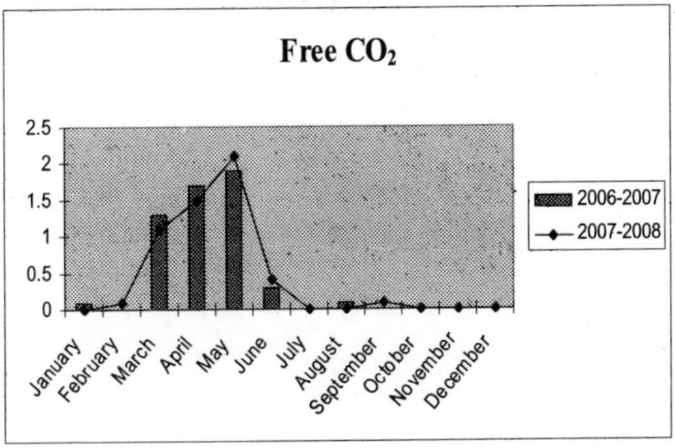

Fig. 6.II.11: Graph showing monthly fluctuation of Free CO_2 at Sampling Site-II in 2007-2008

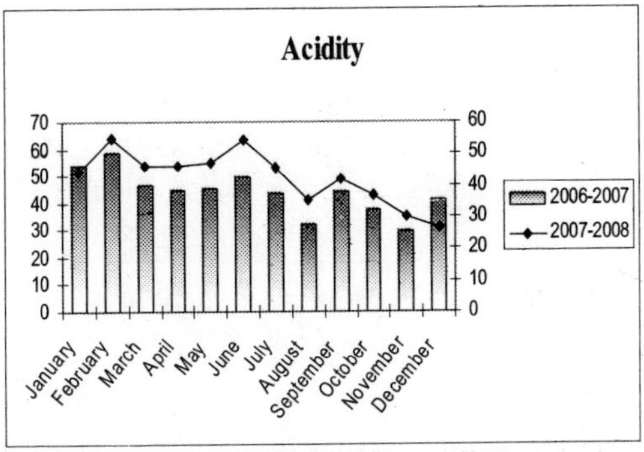

Fig. 6.II.12: Graph showing monthly fluctuation of Acidity at Sampling Site-II in 2007-2008

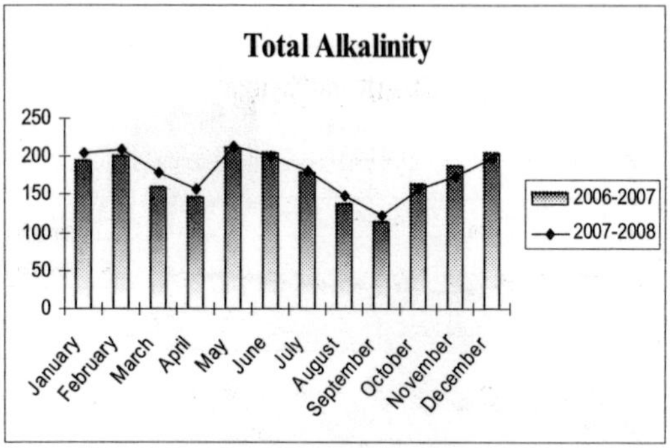

Fig. 6.II.13: Graph showing monthly fluctuation of Total Alkalinity at Sampling Site-II in 2007-2008

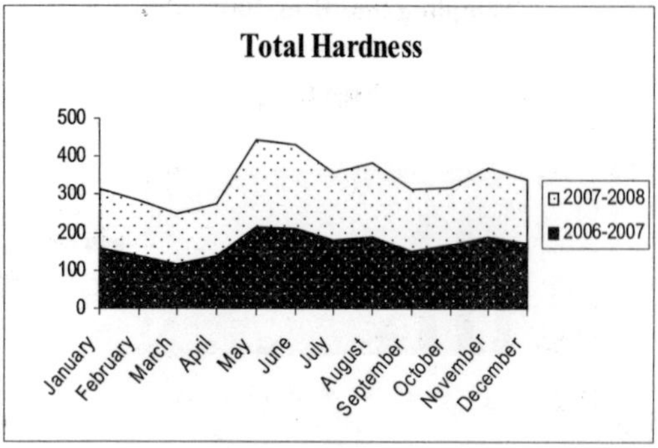

Fig. 6.II.14: Graph showing monthly fluctuation of Total Hardness at Sampling Site-II in 2007-2008

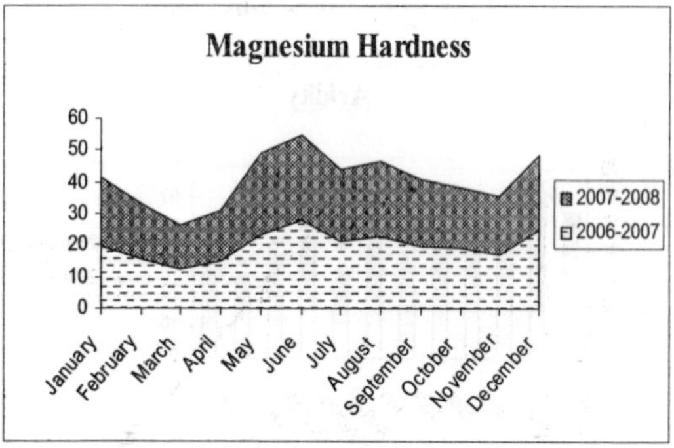

Fig. 6.II.15: Graph showing monthly fluctuation of Calcium Hardness at Sampling Site-II in 2007-2008

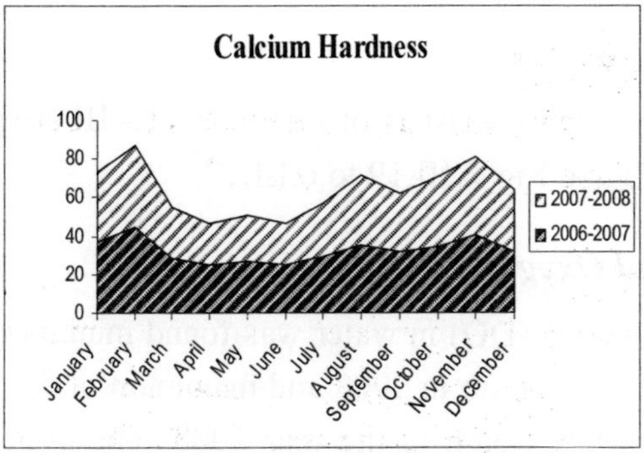

Fig. 6.II.16: Graph showing monthly fluctuation of Magnesium Hardness at Sampling Site-II in 2007-2008

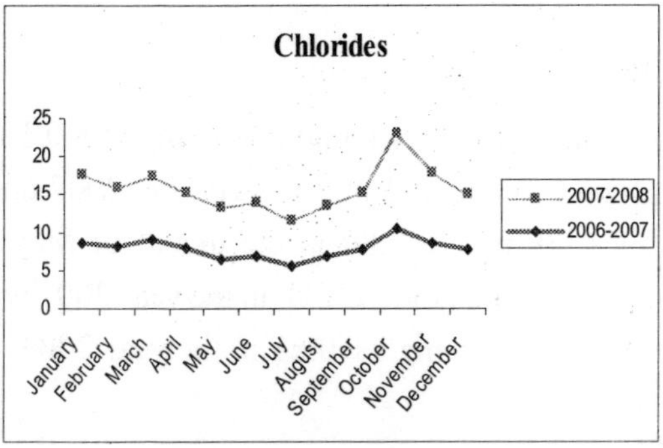

Fig. 6.II.17: Graph showing monthly fluctuation of Chlorides at Sampling Site-II in 2007-2008

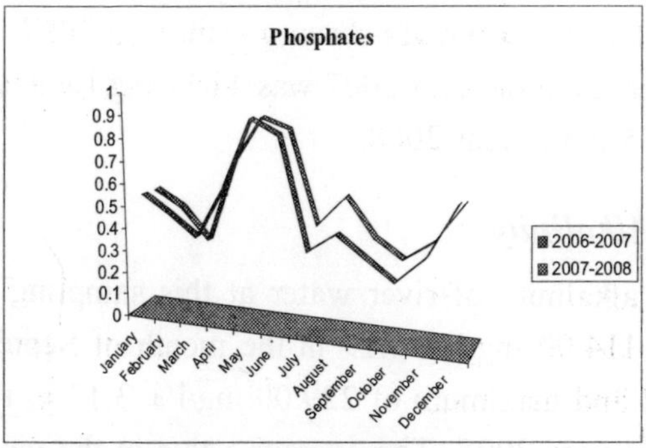

Fig. 6.II.18: Graph showing monthly fluctuation of Phosphates at Sampling Site-II in 2007-2008

Sampling site III

The chemical characteristics of sampling site- III are presented in Table 6.III.3 and Fig. 6.III.19 to 6.III.27.

i. Dissolved Oxygen (DO)

Dissolved oxygen (DO) in water was found minimum 7.4 mg/l ± 0.12 in the May month in 2008 and maximum of 9.9 mg/l ± 0.04 in the December month in the year 2008. The average value of DO in the year 2007 was observed to be 8.73 mg/l ± 0.75 and 8.75 mg/l ± 0.73 in the year 2008.

ii. Free CO_2

Free CO_2 of river water was found minimum with 0.1 mg/l ± 0.01 in the month of January of both the years *i.e.* 2007 and 2008 and maximum of 1.90 mg/l ± 0.89 in the month of May in the year 2008 .The average value of free CO_2 in the year 2007 was observed 0.5 mg/l ± 0.57 and 0.5 mg/l ± 0.61 in the year 2008.

iii. Acidity

The acidity of water was observed minimum of 29.10 mg/l ± 0.25 in November month in the year 2007 and maximum of 59.21 mg/l ± 1.16 in the month of February in the year 2007. The average value observed in the year 2007 was 45.17 mg/l ± 9.02 and 44.68 mg/l ± 7.65 in the year 2008.

iv. Total Alkalinity

The total alkalinity of river water at this sampling station was minimum 114.00 mg/l ± 0.19 in the month of September in the year 2007 and maximum of 229.00 mg/l ± 3.17 in the month of May in the year 2007 .The average value in the year 2007 was 181.29 mg/l ± 36.78 and 181.42 mg/l ± 30.29 in the year 2008.

v. Total Hardness

The total hardness of water was recorded minimum 1085.00 mg/l ± 0.00 in the month of March in the year 2007 and maximum value of 256.00 mg/l ± 0.16 was recorded in the month of May in the year 2008. The average value observed in the year 2007 was 171.17 mg/l ± 40.96 and 177.00 mg/l ± 42.94 in the year 2008.

vi. Calcium Hardness

The calcium hardness in river water at sampling site-III was minimum 23.47 mg/l ± 0.88 in month of June in the year 2008 and maximum of 42.85 mg/l ± 0.24 in February in the same year. The average value in the year 2007 was 32.08 mg/l ± 5.33 and 33.81 mg/l ± 6.24 in the year 2008.

vii. Magnesium Hardness

The magnesium hardness in river water at sampling site-III was found minimum 12.74 mg/l ± 0.05 in the month of March in the year 2007 and maximum 28.37 mg/l ± 0.25 in the month of June in the year 2007. The average value in the year 2007 was observed 20.33 mg/l ± 4.67 and 20.64 mg/l ± 4.16 in the year 2008.

viii. Chlorides

The chlorides in water was minimum 6.10 mg/l ± 0.18 in the month of July in the year 2007 and maximum 9.35 mg/l ± 0.02 in the month of October in the year 2008. The average value of chlorides in the year 2008 was 7.82 mg/l ± 0.94 and 7.92 mg/l ± 1.05 in the year 2008.

ix. Phosphates

At this site the phosphate was recorded to be minimum with 0.30 mg/l ± of October month in year 2008 and maximum 0.87 mg/l ± of May month in year 2007. The annual mean calculated for year 2007 was 0.54 mg/l ± 0.17 and of year 2008 was 0.52 mg/l ± 0.15.

Table 6.III.3: Monthly fluctuation in Chemical parameters of Sampling Site – III during 2007-2008

Months	DO (mg/l)	Free CO_2 (mg/l)	Acidity (mg/l)	Total Alkalinity (mg/l)	Total Hardness (mg/l)	Ca Hardness (mg/l)	Mg Hardness (mg/l)	Chlorides (mg/l)	Phosphates (mg/l)
2007 (Sampling Station – III)									
January	9.3±0.05	0.1±0.01	54.29±2.52	198.0±8.97	143.0±0.05	35.14±6.17	20.37±0.02	8.43±6.13	0.60 ± 0.19
February	9.5±0.00	0.2±0.00	59.21±1.16	205.0±0.40	130.0±0.05	41.73±0.00	16.86±5.26	7.99±0.09	0.54 ± 0.17
March	8.9±0.57	0.7±0.02	48.62±0.23	165.0±0.76	108.0±0.00	32.74±3.01	12.74±0.05	8.76±0.00	0.38 ± 0.18
April	7.8±0.90	1.1±0.60	43.36±0.14	143.0±0.00	140.0±0.18	25.54±0.00	13.20±0.01	7.98±6.71	0.56 ± 0.15
May	7.5±0.20	1.8±0.90	48.90±0.45	229.0±3.17	246.0±0.11	28.10±0.04	24.60±0.00	6.30±0.01	0.87 ± 0.11
June	7.9±0.00	0.5±0.02	53.67±0.08	216.0±0.13	230.0±0.41	24.20±0.23	28.37±0.25	6.98±0.36	0.79 ± 0.09
July	8.4±0.23	0.4±0.01	44.64±0.42	178.5±3.71	189.0±0.04	28.79±0.53	22.74±0.15	6.10±0.18	0.45 ±0.11
August	8.2±0.30	0.2±0.00	32.54±0.23	134.0±0.37	198.0±0.20	34.43±0.58	23.37±0.14	7.92±0.55	0.54 ± 0.10
September	8.9±0.17	0.2±0.00	43.90±0.16	114.0±0.19	159.0±0.90	29.00±0.22	20.83±0.12	8.32±0.17	0.40 ± 0.08
October	9.4±0.22	0.5±0.00	35.91±0.13	176.0±1.00	149.0±0.05	33.61±2.40	19.67±0.03	9.21±1.06	0.31 ± 0.14
November	9.3±0.11	0.7±0.00	29.10±0.25	192.0±7.36	187.0±1.25	39.84±1.12	17.36±0.05	8.31±2.54	0.37 ± 0.17
December	9.7±0.08	0.4±0.00	47.87±1.39	225.0±3.02	175.0±1.05	31.85±0.06	23.84±0.16	7.54±1.00	0.65 ± 0.20
Average±Sd	8.73±0.75	0.5±0.57	45.17±9.02	181.29±36.78	171.17±40.96	32.08±5.33	20.33±4.67	7.82±0.94	0.54±0.17

contd. on next page...

contd...

2008 (Sampling Station – III)									
January	9.5±0.00	0.1±0.01	49.95±0.81	192.0±2.15	151.0±0.05	34.85±2.78	20.56±0.15	8.94±0.02	0.58 ± 0.20
February	9.3±0.02	0.3±0.00	56.82±0.02	200.0±2.56	121.0±0.21	42.85±0.24	16.36±24.57	7.87±0.90	0.50 ± 0.21
March	8.7±0.03	0.6±0.01	44.37±0.16	171.0±2.16	115.0±0.10	29.58±1.64	13.87±0.24	8.25±0.08	0.41 ± 0.19
April	7.9±0.02	1.3±0.05	39.92±0.26	151.0±0.08	142.0±0.02	26.83±0.88	15.96±0.05	7.97±0.02	0.55 ± 0.18
May	7.4±0.12	1.9±0.89	50.35±0.16	223.0±5.68	256.0±0.16	27.43±0.34	25.54±0.78	6.24±0.04	0.82 ±0.17
June	8.2±0.15	0.4±0.12	56.22±0.02	208.0±0.06	232.0±0.24	23.47±0.88	27.98±0.05	6.64±0.16	0.74 ± 0.14
July	8.5±0.05	0.5±0.00	45.15±0.05	182.0±0.18	198.0±0.12	33.85±2.38	21.10±0.15	6.21±0.00	0.43 ± 0.15
August	8.3±0.06	0.4±0.00	39.27±0.20	143.0±0.16	209.0±0.16	40.63±1.85	23.74±0.19	7.92±0.32	0.52 ± 0.11
September	8.8±0.21	0.5±0.01	45.78±0.16	122.0±5.25	169.0±0.25	34.87±3.31	20.95±1.07	8.93±0.50	0.38 ± 0.09
October	9.4±0.24	0.7±0.00	37.73±0.21	183.0±5.68	158.0±6.21	37.84±1.43	19.58±29.97	9.35±0.02	0.30 ± 0.10
November	9.1±0.09	0.6±0.01	31.84±0.25	187.0±8.61	192.0±0.83	41.74±0.63	18.34±3.34	8.56±0.70	0.39 ± 0.12
December	9.9±0.04	0.4±0.00	38.70±5.25	215.0±2.26	181.0±0.05	31.74±4.82	23.75±0.00	8.12±0.06	0.62 ± 0.08
Average±Sd	8.75±0.73	0.5±0.61	44.68±7.65	181.42±30.29	177.00±42.94	33.81±6.24	20.64±4.16	7.92±1.05	0.52± 0.15

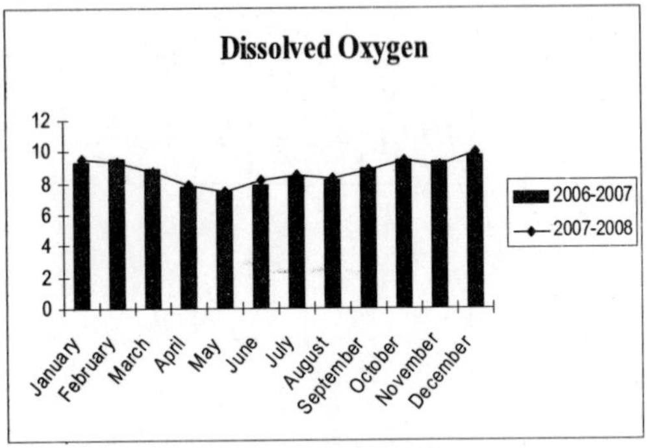

Fig. 6.III.19: Graph showing monthly fluctuation of Dissolved Oxygen at Sampling Site-III in 2007-2008

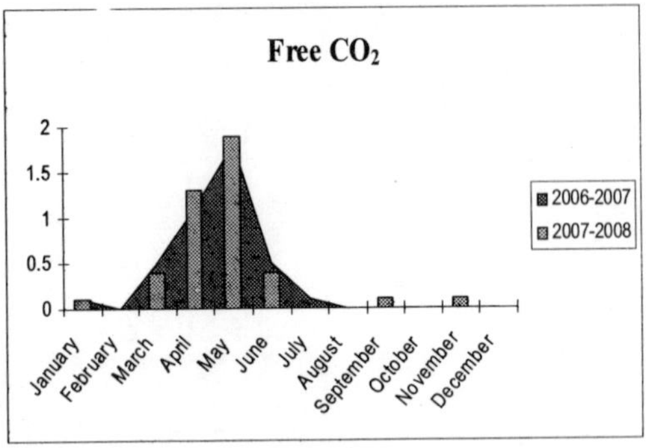

Fig. 6.III.20: Graph showing monthly fluctuation of Free CO_2 at Sampling Site-III in 2007-2008

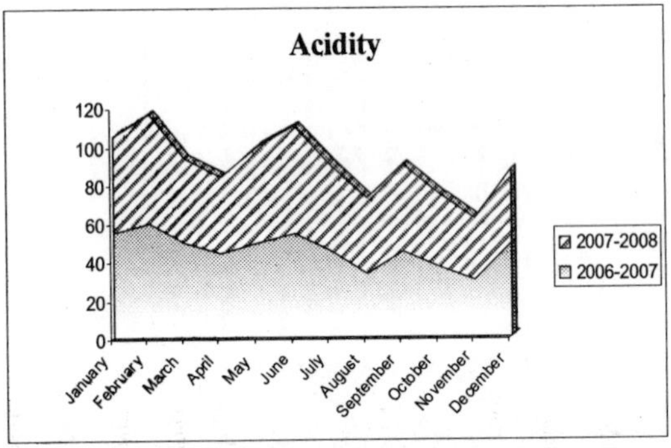

Fig. 6.III.21: Graph showing monthly fluctuation of Acidity at Sampling Site-III in 2007-2008

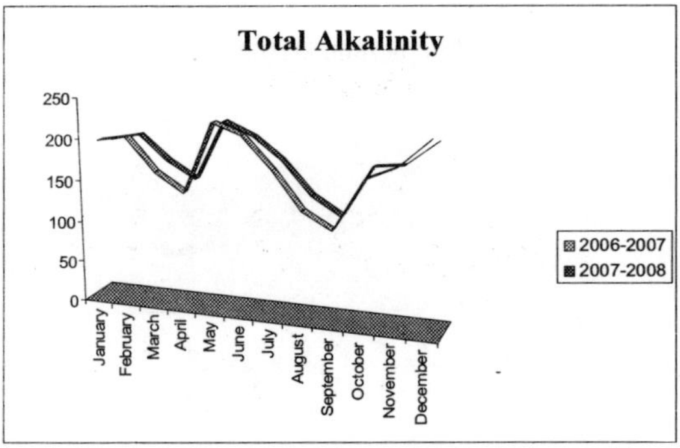

Fig. 6.III.22: Graph showing monthly fluctuation of Total Alkalinity at Sampling Site-III in 2007-2008

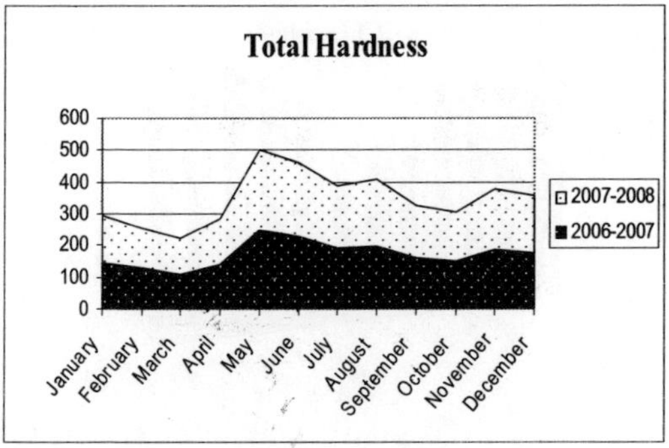

Fig. 6.III.23: Graph showing monthly fluctuation of Total Hardness at Sampling Site-III in 2007-2008

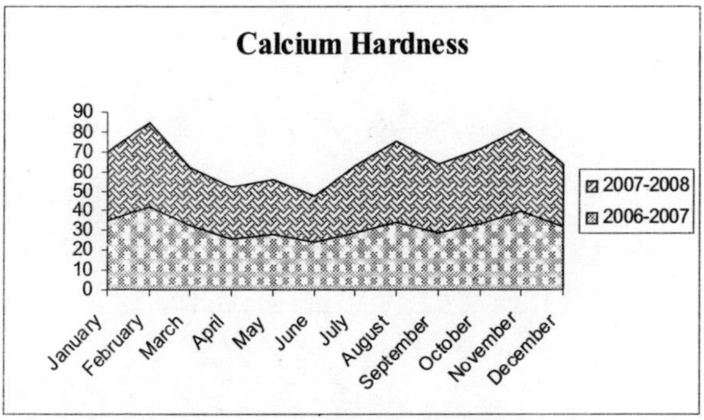

Fig. 6.III.24: Graph showing monthly fluctuation of Calcium Hardness at Sampling Site-III in 2007-2008

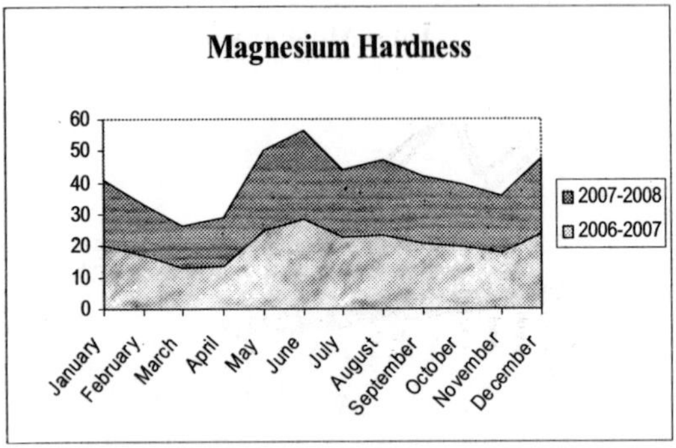

Fig. 6.III.25: Graph showing monthly fluctuation of Magnesium Hardness at Sampling Site-III in 2007-2008

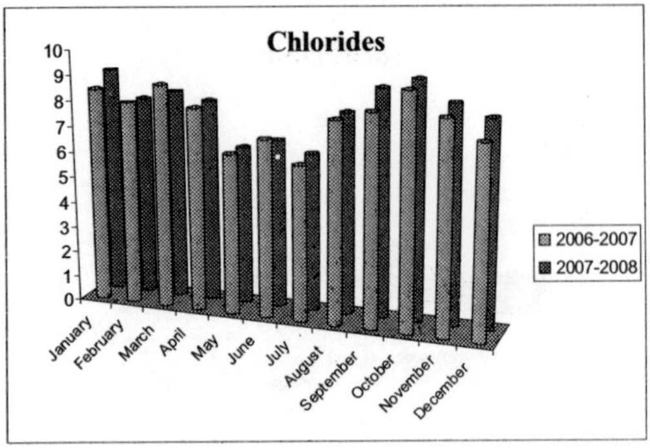

Fig. 6.III.26: Graph showing monthly fluctuation of Chlorides at Sampling Site-III in 2007-2008

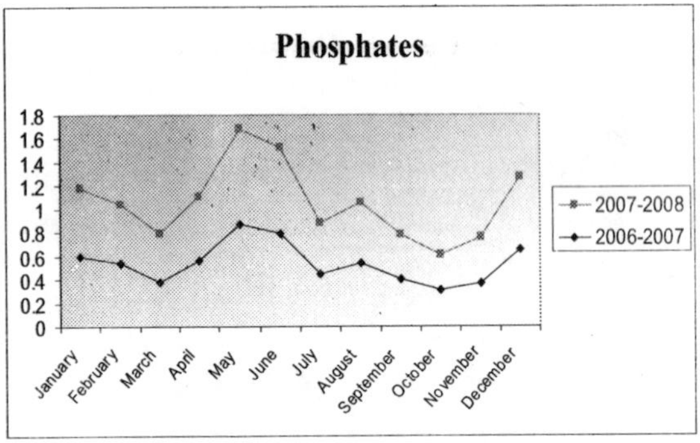

Fig. 6.III.27: Graph showing monthly fluctuation of Phosphates at Sampling Site-III in 2007-2008

CHAPTER-7

BIODIVERSITY ANALYSIS

A study on the biodiversity of the Henwal River includes the primary producers like phytoplankton, Consumers like zooplankton, secondary and tertiary consumers like fishes and benthic fauna. Site wise details of the observations of biological parameters are mentioned in tables (Table 7. I. A.1 to 7.I. A. 9) and (Fig. 7.I. A.1 to 7.I. A.18)

A. Planktonic Diversity

Sampling Site I

(a) Plankton

Quantitative analysis of plankton at sampling site I revealed that total planktonic number per liter of water was observed maximum 1385.72 unit/l ± 6.30 in the month of December in the year 2007 and was minimum was 838.74 unit/l ± 13.44 in July month of the 2008. The average value of plankton in 2007 is 1116.23 ± 179.25 and in 2008 was 1115.92 ± 189.08 (Table 7.I. A.1 to Table 7.I. A.3 and Figs. 7.I.A.1 to 7.I.A.6)

1. Phytoplankton:

Phytoplankton were noted lowest of 585.34 unit/l ± 4.35 in the month of July of 2008 and highest of 1067.12 unit/l ± 1.99 in

December month of year 2007 with the average cells number 822.14 unit/l ± 163.54 in 2007 and 823.69 unit/l ± 167.26 in 2008.

In phytoplankton Blue green algae was dominant and class Diatoms was found least in both the years of study. At this sampling site Diatoms was observed maximum in the month of March of 2008 (304.46 unit/l ±0.61) and minimum in August month (178.52 unit/l ± 2.23) of 2007-2008 with average count of cells in 2007 was 231.39 ± 44.00 unit/l and 226.25 ± 40.02 unit/l in 2007-2008. Green algae were found to be highest in the month of April 2008 (391.82 unit/l ± 1.20) and lowest in 2008 in the month of September (156.43unit/l ± 1.29). The average count of Green algae in 2007 was 257.05 unit/l ± 72.79 and 265.02 ± 76.42 unit/l in 2008. Blue green algae showed optimum peak during the month of February in 2008 (489.49 unit/l ± 0.95) and least during the month of July of 2008 (215.33 unit/l ± 1.57). In 2007 the average value of Blue green algae was 333.70 unit/l ± 76.99 and in 2008 it was 332.42 unit/l ± 85.58.

2. Zooplankton:

Zooplankton were observed minimum in July of 2008 (253.03 unit/l ± 6.66) and maximum in 2008 for the month of December (342.93 unit/l 5.01) and the average value 294.09 unit/l ± 27.16 in 2007 and in 2007-2008 the average was 292.22 unit/l ± 27.83.

At this site *Protozoa* was maximum in the month of January in 2007 (107.30 unit/l ± 0.17) and minimum was found 61.25 unit/l ± 1.11 in June month (2007). *Rotifera* was found to be highest (82.35 unit/l ± 0.25) in the month of November 2007 and lowest (49.14 unit/l ± 2.57) in the month of July of the same year (2007). *Cladocera* was observed in their optimum peak during 2007 January (88.29 unit/l ± 0.69) and was lowest during the month of

August (49.98 unit/l ± 1.09) in the year 2007. *Copepod* was maximum (58.91 unit/l ± 1.27) in 2008 January and minimum was found in the month of July 2008 (30.75 unit/l ± 0.77). *Ostracods* sp. was observed in their optimum peak during 2008 December (46.58 unit/l ± 1.57) and was lowest during the month of July (25.49 unit/l ± 0.31) in the year 2007.

The average value revealed that *Protozoa* was dominating in the zooplankton. In 2007 (87.13 unit/l ± 14.62) followed by *Cladocera* (68.45 unit/l ± 10.83), *Rotifera* (61.67 unit/l ± 8.65), *Copepod* (44.39 unit/l ± 7.94), *Ostracods* sp. (32.45 unit/l ± 4.90). In second year 2008 *Protozoa* (86.49 unit/l ± 10.93) found in strength followed by *Cladocera* (66.54 unit/l ± 9.21), *Rotifera* (61.12 unit/l ± 8.10), *Copepod* (43.65 unit/l ± 8.01) and *Ostracods* sp (34.43 unit/l ± 4.99).

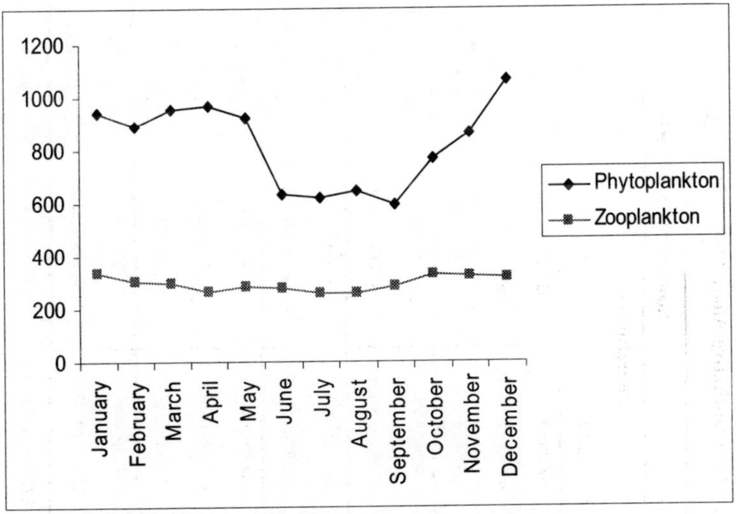

Fig. 7.I.A.1: Graph showing the monthly variation in Plankton at Sampling Site-I (2007)

Table 7.I.A.1: Quantitative Analysis of the Plankton of the Henwal River During 2006-2008 at sampling station I

Months	Number of Plankton per litre of water					
	Phytoplankton		Zooplankton		Total Plankton	
	2007	2008	2007	2008	2007	2008
January	940.86±1.00	930.81±0.93	333.04±2.79	328.66±3.19	1273.90±3.79	1259.47±4.12
February	889.01±1.16	1004.30±2.19	306.35±3.28	299.96±3.33	1195.36±4.44	1304.26±5.52
March	957.83±1.11	932.29±2.94	294.27±3.36	306.17±3.64	1252.10±4.47	1238.46±6.58
April	968.40±0.82	931.68±3.41	266.01±4.88	266.50±4.1	1234.41±5.70	1198.18±7.51
May	922.37±1.83	905.89±3.79	285.52±4.55	289.84±5.34	1207.89±6.38	1195.73±9.13
June	633.78±1.69	654.27±5.40	276.56±3.75	289.55±4.3	910.34±5.44	943.82±9.70
July	616.93±1.86	585.34±4.35	259.71±5.80	253.03±6.66	876.64±7.21	838.74±13.44
August	644.35±1.40	621.91±5.20	256.40±6.37	263.98±5.45	900.75±7.77	885.89±10.65
September	594.98±1.41	601.95±6.78	282.55±6.08	259.38±5.09	877.53±7.94	861.33±9.44
October	765.74±2.26	764.75±1.37	326.85±4.78	299.26±5.46	1092.59±6.77	1064.01±6.70
November	864.29±2.24	910.66±1.22	323.20±4.92	307.42±5.05	1187.49±7.16	1218.08±6.27
December	1067.12±1.99	1040.45±1.24	318.60±4.04	342.93±5.01	1385.72±6.30	1383.38±3.38
Average ±Sd	822.14±163.54	823.69±167.26	294.09±27.16	292.22±27.83	1116.23±179.25	1115.92±189.08

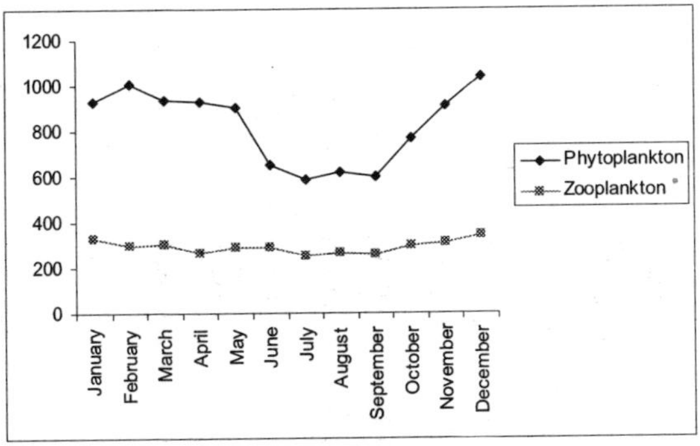

Fig. 7.I.A.2: Graph showing the monthly variation in Plankton at Sampling Site-I (2008)

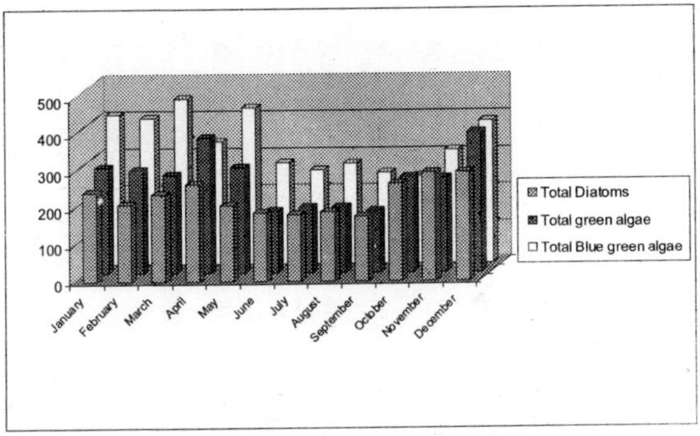

Fig. 7.I.A.3: Graph showing the monthly variation in different Groups of phytoplankton at Sampling Site-I (2007)

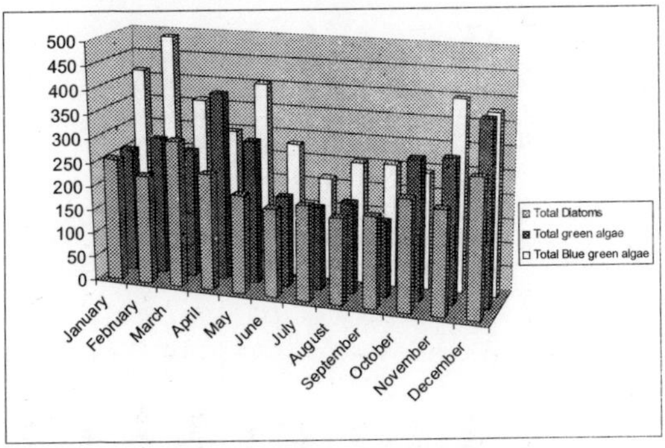

Fig. 7.I.A.4 Graph showing the monthly variation in different Groups of phytoplankton at Sampling Site-I (2008)

Table 7.I.A.2: Number of different Group among the Phytoplankton of the Henwal River During 2007-2008 at sampling station I

Months	Total Phytoplankton/ Liter of water		Total Diatoms / Litre of water		Total green algae/ Litre of water		Total Blue green algae/ Litre of water	
	2007	**2008**	**2007**	**2008**	**2007**	**2008**	**2007**	**2008**
January	940.86±1.00	930.81±0.93	245.12±0.13	256.67±0.52	287.46±0.55	259.82±0.29	408.28±0.32	414.32±0.12
February	889.01±1.16	1004.30±2.19	211.34±0.21	226.54±0.26	279.29±0.68	288.27±0.98	398.38±0.27	489.49±0.95
March	957.83±1.11	932.29±2.94	238.27±0.33	304.46±0.61	268.37±0.29	268.28±1.11	451.19±0.49	359.55±1.22
April	968.40±0.82	931.68±3.41	266.67±0.38	241.78±1.18	368.51±0.09	391.82±1.20	333.22±0.35	298.08±1.03
May	922.37±1.83	905.89±3.79	208.13±0.43	204.88±2.11	288.11±0.38	298.90±1.31	426.13±1.02	402.11±0.37
June	633.78±1.69	654.27±5.40	189.33±0.28	184.09±3.55	168.05±0.51	188.81±1.28	276.4±0.90	281.37±0.57
July	616.93±1.86	585.34±4.35	183.20±0.21	197.19±1.21	178.57±0.57	172.82±1.57	255.16±1.08	215.33±1.57
August	644.35±1.40	621.91±5.20	191.27±0.52	178.52±2.23	179.80±0.31	189.60±1.22	273.28±0.57	253.79±1.75
September	594.98±1.41	601.95±6.78	179.28±0.42	189.24±3.41	168.61±0.70	156.43±1.29	247.09±0.29	256.28±2.08
October	765.74±2.26	764.75±1.37	267.40±0.23	231.22±0.32	259.77±0.92	289.55±0.03	238.57±1.11	243.98±1.02
November	864.29±2.24	910.66±1.22	209.05±0.11	215.34±0.03	256.52±1.11	296.76±0.03	309.43±1.02	398.56±1.16
December	1067.12±1.99	1040.45±1.24	298.34±0.22	285.07±0.21	381.49±0.36	379.19±0.05	387.29±1.41	376.19±0.98
Average±Sd	822.14±163.54	823.69±167.26	231.39±44.00	226.25±40.02	257.05±72.79	265.02±76.42	333.70±76.99	332.42±85.58

Table 7.I.A.3: Number of Different Genera among Zooplankton in the Henwal River During 2007-2008 at sampling station I

Months	*Protozoa*		*Rotifera*		*Cladocera*		*Copepod*		*Ostracods*	
	2007	2008	2007	2008	2007	2008	2007	2008	2007	2008
January	107.30±0.17	102.50±0.24	59.25±0.57	59.38±0.58	88.29±0.69	69.28±0.28	48.83±0.38	58.91±1.27	29.37±1.29	38.59±1.28
February	97.95±0.23	88.52±0.35	62.26±0.34	60.25±0.39	76.28±0.85	79.18±0.82	35.27±0.85	36.72±0.72	34.59±1.01	35.29±1.05
March	99.25±0.53	90.15±0.35	57.67±0.79	56.27±0.58	68.57±0.38	79.19±0.85	31.24±0.81	45.27±0.81	37.54±0.73	35.29±1.09
April	76.54±0.15	77.25±0.24	51.29±1.58	56.24±0.99	62.82±0.82	61.85±0.42	40.14±1.12	38.79±1.08	35.22±1.21	32.37±1.37
May	85.55±1.04	89.85±1.05	59.85±0.59	61.45±0.38	60.92±0.27	58.76±1.08	51.02±0.76	50.19±1.12	28.18±1.89	29.59±1.71
June	61.25±1.11	71.05±1.01	66.08±0.02	61.29±0.38	72.16±0.52	75.29±0.94	49.78±1.02	46.23±0.83	27.29±1.08	35.69±1.14
July	68.57±1.25	70.05±1.35	49.14±2.57	59.83±1.57	61.23±0.85	60.48±1.81	55.28±1.59	30.75±0.77	25.49±0.31	31.92±0.66
August	73.25±1.65	79.25±1.55	58.59±1.57	53.48±1.04	49.98±1.09	55.28±0.90	38.85±0.58	39.19±0.43	35.73±1.48	36.78±1.53
September	88.58±0.28	84.47±0.38	63.98±1.86	51.82±1.98	57.84±1.19	55.08±1.15	41.07±0.82	40.08±0.23	31.08±1.22	27.93±1.35
October	92.89±0.68	96.28±0.88	69.88±1.58	58.97±1.25	83.05±0.81	77.85±1.09	38.28±0.28	36.59±0.67	42.75±1.43	29.57±1.87
November	89.39±0.89	85.25±0.89	82.35±0.25	76.85±0.31	72.08±1.01	62.89±0.98	47.71±1.01	48.89±1.38	31.67±1.76	33.54±1.49
December	105.01±0.21	103.22±0.58	59.68±0.57	77.58±0.68	68.19±1.08	63.38±0.85	55.25±0.76	52.17±1.03	30.47±1.42	46.58±1.57
Average ±Sd	**87.13±14.62**	**86.49±10.93**	**61.67±8.65**	**61.12±8.10**	**68.45±10.83**	**66.54±9.21**	**44.39±7.94**	**43.65±8.01**	**32.45±4.90**	**34.43±4.99**

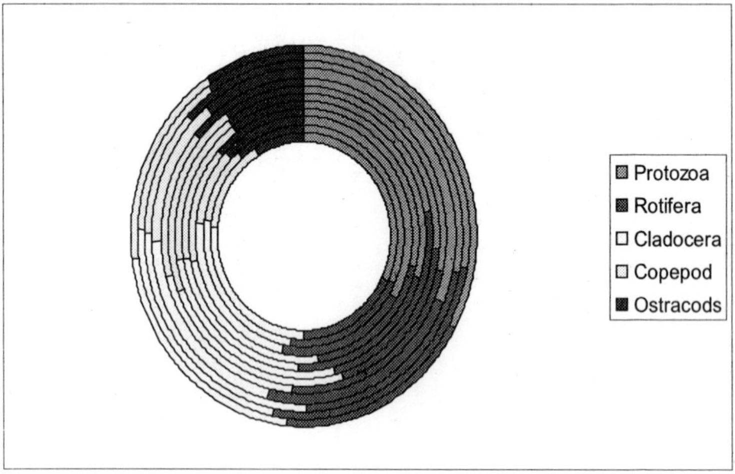

Fig. 7.I.A.5:Graph showing the monthly variation in different Genera of Zooplankton at Sampling Site-I (2007)

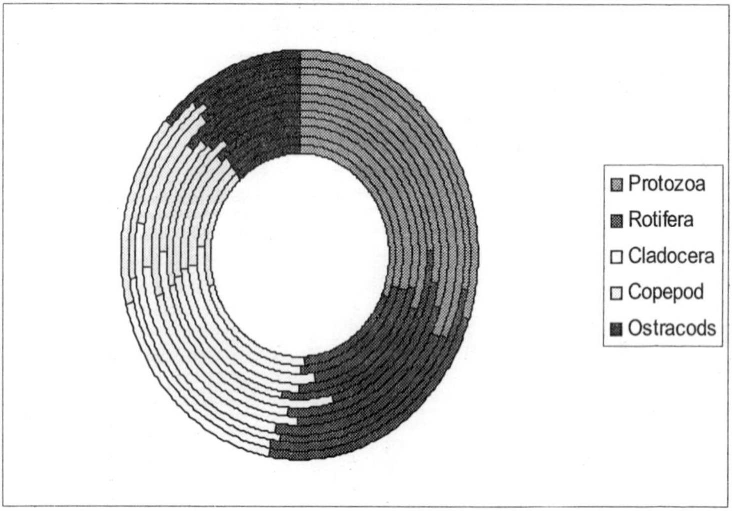

Fig. 7.I.A.6: Graph showing the monthly variation in different Genera of Zooplankton at Sampling Site-I (2008)

Sampling Site II

(a) Plankton:

Quantitative analysis of plankton at sampling site-II revealed that the total planktonic number per liter of water was observed maximum 1322.94 unit/l ± 6.37 in the month of February in the year 2007 and minimum was 695.28 unit/l ± 11.90 in June month of the 2008. The average value of plankton in 2007 is 1053.52 ± 225.84 and in 2008 was 1078.69 ± 234.17 (Table 7.II.A.4 to Table 7.II.A.6 and Figs. 7.II.A.7 to 7.II.A.12)

1. Phytoplankton:

Phytoplankton was noted lowest of 401.78 unit/l ± 516 in the month of June of 2008 and was highest 1022.36 unit/l ± 1.17 in February month of 2007 with the average cells number 777.61 unit/l ± 204.67 in 2007 and 789.13 unit/l ± 222.19 in 2008.

In phytoplankton Blue green algae were dominated and class Diatoms was found least in both the years of study. At this sampling site Diatoms was observed maximum in the month of March of 2008 (324.49 unit/l ± 0.21) and minimum in August month (108.51 unit/l ± 2.83) of 2008 with average count of cells in 2007 was 201.65 ± 53.73 unit/l and 203.13 ± 66.08 unit/l in 2008. Green algae were found to be highest in the month of October 2008 (399.81 unit/l ± 0.15) and lowest in 2007 in the month of July (128.36 unit/l ± 1.70). The average count of Green algae in 2007 was 252.76 unit/l ± 79.84 and 264.49 ± 78.85 unit/l in 2008. Blue green algae showed optimum peak during the month of February in 2007 (491.81 unit/l ± 0.89) and was least during the month of June 2008 (203.19 unit/l ± 1.31). In 2007 the average value of Blue green algae was 320.69 unit/l ± 97.35 and in 2008 it was 268.57 unit/l ± 98.37.

2. Zooplankton:

Zooplankton was observed minimum in July of 2007 (199.06 unit/l ± 6.85) and maximum in 2008 for the month of December (342.87 unit/l 6.34) and the average value of 275.91 unit/l ± 35.36 in 2007 and in 2008 the average was 289.56 unit/l ± 28.78.

At this site *Protozoa* was maximum in the month of January in 2008 (131.05 unit/l ± 0.13) and minimum was found 39.81 unit/l ± 1.03 in July month of 2007. *Rotifera* was found to be highest in the month of November 2007 (83.15 unit/l ± 0.35) and lowest (44.14 unit/l ± 2.42) in the month of July of the year 2007. *Cladocera* was observed in their optimum peak during 2007 October (88.01 unit/l ± 0.91) and is lowest during the month of August (49.99 unit/l ± 1.49) in the year 2007. *Copepod* was maximum (59.07 unit/l ± 0.66) in 2007 May and minimum was found in the month of July 2007 (30.20 unit/l ± 1.59). *Ostracods* sp. was observed in their optimum peak during 2007 December (48.77 unit/l ± 1.45) and is lowest during the month of July (20.56 unit/l ± 0.62) in the year 2008.

The average value revealed that *Protozoa* was dominating in the zooplankton. In 2007 (71.95 unit/l ± 22.58) followed by *Cladocera* (68.36 unit/l ± 11.30), *Rotifera* (59.54 unit/l ± 9.57), *Copepod* (44.56 unit/l ± 8.24), *Ostracods* sp. (31.51 unit/l ± 7.35). In second year 2008 *Protozoa* (88.64 unit/l ± 16.32) found in strength followed by *Cladocera* (65.26 unit/l ± 10.25), *Rotifera* (60.64 unit/l ± 9.11) *Copepod* (41.53 unit/l ± 8.43) and *Ostracods sp* (33.48 unit/l ± 6.13).

Table 7.II.A.4: Quantitative Analysis of the Plankton of the Henwal River During 2007-2008 at sampling station II

Months	Number of Plankton per litre of water				Total Plankton	
	Phytoplankton		Zooplankton			
	2007	2008	2007	2008	2007	2008
January	932.95±2.1	939.18±2.63	294.18±3.04	317.49±3.27	1227.13±5.14	1256.67±5.90
February	1022.36±1.17	1013.45±3.69	300.58±5.2	290.85±2.92	1322.94±6.37	1304.30±6.61
March	875.02±1.95	889.88±2.69	295.74±6.03	278.38±4.61	1170.76±7.98	1168.26±7.30
April	848.61±1.01	935.72±3.54	284.08±4.84	284.50±4.99	1132.69±5.85	1220.22±8.53
May	973.74±3.05	987.05±3.96	261.00±4.18	280.03±5.5	1234.74±7.23	1267.08±9.46
June	544.26±3.20	401.78±5.16	273.66±5.02	293.50±6.74	817.92±8.22	695.28±11.90
July	525.11±3.57	561.33±4.80	199.06±6.85	249.72±6.84	724.17±10.42	811.05±11.64
August	504.96±1.97	487.20±5.32	228.05±6.81	240.14±5.69	733.01±8.78	727.34±11.01
September	527.09±1.94	580.89±4.33	264.64±6.58	287.01±5.45	791.73±8.52	867.90±9.78
October	1007.61±3.36	1003.41±1.43	270.31±5.37	286.23±4.77	1277.92±8.73	1289.64±6.20
November	863.70±3.05	915.04±2.54	326.49±5.39	323.94±4.45	1190.19±8.44	1238.98±6.99
December	705.95±3.37	754.68±2.41	313.14±4.23	342.87±6.34	1019.09±7.60	1097.55±8.75
Average ±Sd	777.61±204.67	789.13±222.19	275.91±35.56	289.56±28.78	1053.52±225.84	1078.69±234.17

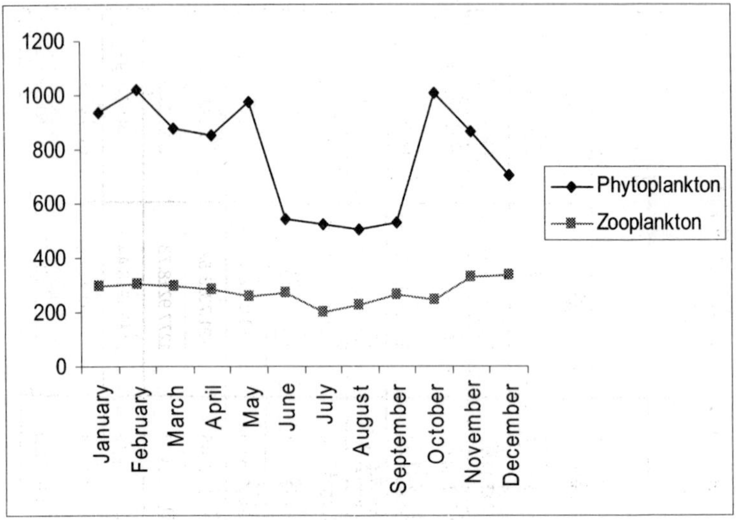

Fig. 7.II.A.7: Graph showing the monthly variation in Plankton at Sampling Site-II (2007)

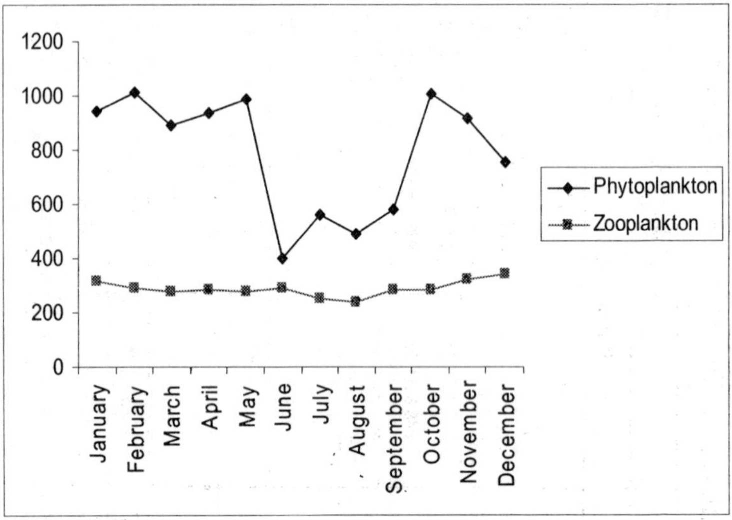

Fig. 7.II.A.8: Graph showing the monthly variation in Plankton at Sampling Site-II (2008)

Table 7.II.A.5: Number of different Group among the Phytoplankton of the Henwal River During 2007-2008 at sampling station II

Months	Total Phytoplankton/ Liter of water		Total Diatoms / Litre of water		Total green algae/ Litre of water		Total Blue green algae/ Litre of water	
	2007	**2008**	**2007**	**2008**	**2007**	**2008**	**2007**	**2008**
January	932.95±2.1	939.18±2.63	225.28±0.23	236.46±1.12	289.16±0.95	299.90±0.39	418.51±0.92	402.82±1.12
February	1022.36±1.17	1013.45±3.69	231.37±0.23	266.76±1.06	299.18±0.69	258.81±1.58	491.81±0.89	487.88±1.05
March	875.02±1.95	889.88±2.69	248.51±0.37	324.49±0.21	288.27±0.69	208.46±1.16	338.24±0.25	356.93±1.32
April	848.61±1.01	935.72±3.54	206.62±0.27	261.85±1.13	338.38±0.29	390.15±1.25	303.61±0.45	283.72±1.51
May	973.74±3.05	987.05±3.96	208.85±0.55	274.19±1.21	298.46±1.38	294.49±1.39	466.43±1.12	418.37±1.36
June	544.26±3.20	401.78±5.16	139.29±0.37	164..38±2.45	198.49±1.61	198.59±1.20	206.48±1.22	203.19±1.31
July	525.11±3.57	561.33±4.80	119.22±0.65	119.82±2.51	128.36±1.70	186.84±1.28	277.53±1.22	254.67±1.01
August	504.96±1.97	487.20±5.32	151.73±0.19	108.51±2.83	149.35±0.71	169.75±1.24	203.88±1.07	208.94±1.25
September	527.09±1.94	580.89±4.33	133.18±0.37	187.42±1.21	168.15±0.56	178.92±1.56	225.76±1.01	214.55±1.56
October	1007.61±3.36	1003.41±1.43	248.46±0.19	255.11±0.29	391.23±1.36	399.81±0.15	367.92±1.81	348.49±0.99
November	863.70±3.05	915.04±2.54	289.52±0.42	235.09±0.53	254.19±1.31	298.67±0.83	319.99±1.32	381.28±1.18
December	705.95±3.37	754.68±2.41	217.82±0.27	261.73±0.36	229.94±1.92	289.49±0.43	258.19±1.18	203.46±1.62
Average±Sd	777.61±204.67	789.13±222.19	201.65±53.73	203.13±66.08	252.76±79.84	264.49±78.85	320.69±97.35	268.57±98.37

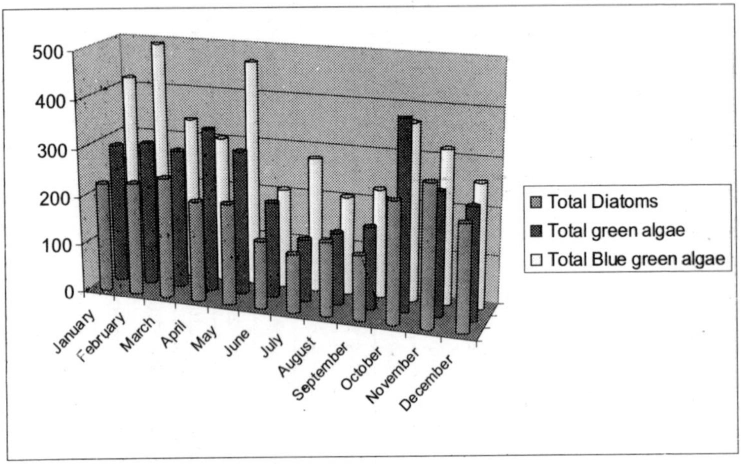

Fig. 7.II.A.9: Graph showing the monthly variation in different Groups of phytoplankton at Sampling Site-II (2007)

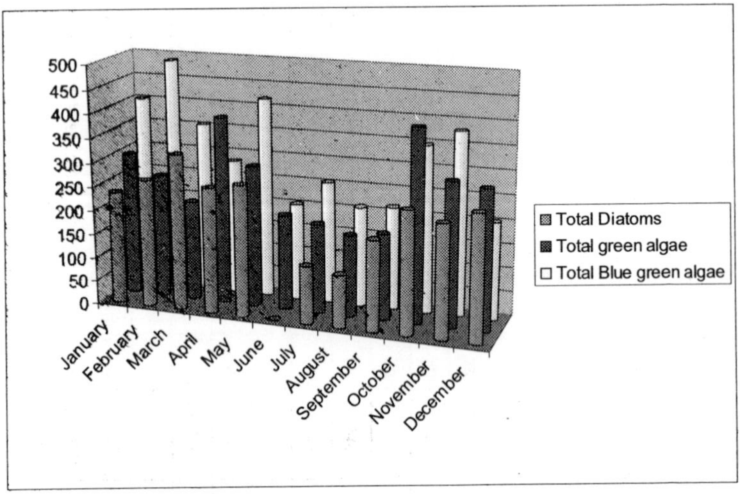

Fig. 7.II.A.10: Graph showing the monthly variation in different Groups of phytoplankton at Sampling Site-II (2008)

Table 7.II.A.6: Number of Different Genera among Zooplankton in the Henwal River During 2007-2008 at sampling station II

Months	*Protozoa*		*Rotifera*		*Cladocera*		*Copepod*		*Ostracods*	
	2007	2008	2007	2008	2007	2008	2007	2008	2007	2008
January	103.19±0.57	131.05±0.13	50.26±0.52	50.33±0.50	64.57±0.36	60.28±0.21	46.83±0.39	41.27±1.21	29.33±1.20	34.56±1.22
February	98.28±1.25	84.52±0.56	60.26±0.44	62.25±0.37	71.26±0.75	75.15±0.52	39.26±1.35	32.70±0.32	31.52±1.41	36.23±1.15
March	79.25±0.98	93.11±0.28	58.65±1.27	51.26±0.53	84.22±1.65	72.09±1.05	40.09±1.41	31.72±1.71	33.53±0.72	30.20±1.04
April	88.67±0.37	75.37±0.54	53.22±1.76	66.24±0.92	67.81±0.32	66.81±1.02	44.16±1.12	37.76±1.18	30.22±1.27	38.32±1.33
May	59.97±0.58	86.18±1.23	60.85±0.57	62.45±0.32	59.95±0.57	51.73±1.02	59.07±0.66	52.17±1.16	21.16±1.80	27.50±1.77
June	67.51±0.88	73.05±1.89	60.02±0.42	64.29±0.88	78.14±1.52	79.24±1.94	42.79±1.02	40.26±0.87	25.20±1.18	36.66±1.16
July	39.81±1.03	86.27±0.18	44.14±2.42	50.82±1.58	54.83±1.39	51.04±1.13	30.20±1.59	41.03±1.23	30.08±1.02	20.56±0.62
August	43.84±1.80	72.27±1.65	56.59±1.52	53.48±1.14	49.99±1.49	50.29±0.91	37.87±1.58	37.14±0.43	39.76±1.42	26.96±1.56
September	55.12±0.83	77.11±1.49	65.58±1.37	58.83±1.56	64.27±1.75	61.46±1.91	56.20±0.72	56.90±1.26	23.47±1.31	32.71±1.33
October	46.58±0.82	86.28±0.37	65.08±1.51	55.57±1.21	88.01±0.91	76.86±1.08	39.23±0.68	30.50±0.57	31.41±1.43	37.02±1.54
November	89.88±0.59	95.10±0.49	83.15±0.35	73.65±0.30	73.05±1.51	68.87±0.90	46.72±1.21	46.81±1.36	33.69±1.73	39.51±1.40
December	91.27±0.39	103.32±0.51	56.67±0.27	78.56±1.68	64.17±1.18	69.30±1.25	52.26±0.96	50.12±1.07	48.77±1.45	41.57±1.83
Average ± Sd	71.95±22.58	88.64±16.32	59.54±9.57	60.64±9.11	68.36±11.30	65.26±10.25	44.56±8.24	41.53±8.43	31.51±7.35	33.48±6.13

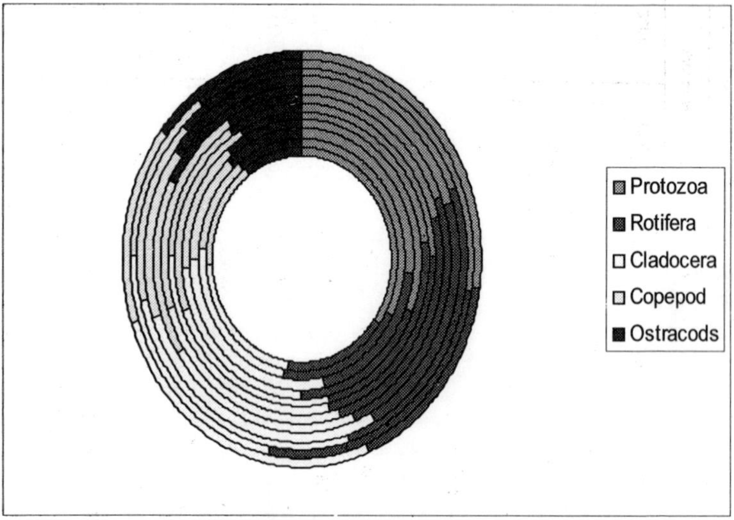

Fig. 7.II.A.11:Graph showing the monthly variation in different Genera of Zooplankton at Sampling Site-II (2007)

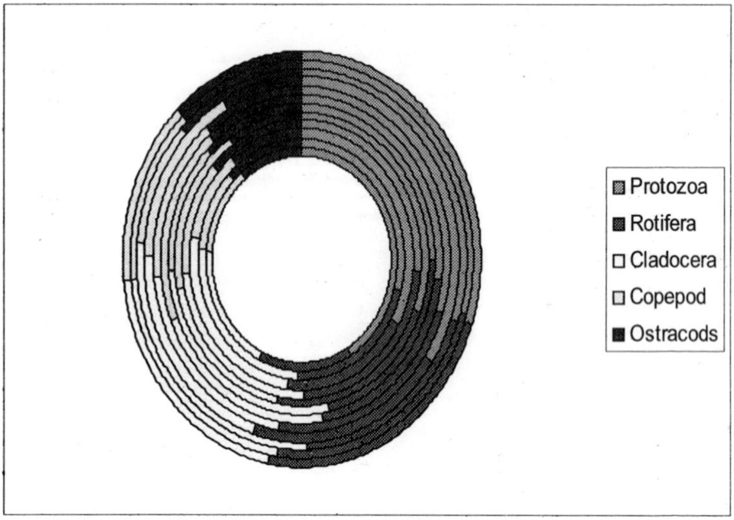

Fig. 7.II.A.12:Graph showing the monthly variation in different Genera of Zooplankton at Sampling Site-II (2008)

Sampling Site III

(a) Plankton:

Quantitative analysis of plankton at sampling site III revealed that the total planktonic number per liter of water was observed maximum 1315.74 unit/l ± 9.04 in the month of February in the year 2008 and minimum was 772.15 unit/l ± 10.35 in August month of the 2007. The average value of plankton in 2007 is 1070.62 ± 175.40 and in 2008 was 1135.73 ± 152.04. (Table 7.III.A.7 to Table 7.III.A.9 and Figs. 7.III.A.13 to 7.III.A.18)

1. Phytoplankton:

Phytoplankton was noted lowest of 555.09 unit/l ± 1.97 in the month of August of 2007 and highest of 1024.89 unit/l ± 3.69 in February month of 2008 with the average cells number 793.36 unit/l ± 148.06 in 2007 and 846.29 unit/l ± 132.32 in 2008.

In phytoplankton Blue green algae were dominated and class Diatoms was found least in both the years of study. At this sampling site Diatoms was observed maximum in the month of December of 2008 (298.75 unit/l ±0.36) and minimum in July month (132.12 unit/l ± 0.65) of 2007 with average count of cells in 2007 was 207.08 ± 44.58 unit/l and 228.14 ± 44.32 unit/l in 2008. Green algae were found to be highest in the month of April 2007 (317.25 unit/l ± 0.29) and lowest in 2007 in the month of August (161.53 unit/l ± 0.71). The average count of Green algae in 2007 was 255.51 unit/l ± 55.03 and 232.70 ± 54.69 unit/l in 2008. Blue green algae showed optimum peak during the month of February in 2008 (457.75 unit/l ± 1.05) and was least during the month of August of 2007 (244.32 unit/l ± 1.07). In 2007 the average value of Blue green algae was 330.77 unit/l ± 67.79 and in 2008 it was 321.29 unit/l ± 54.68.

2. Zooplankton:

Zooplankton was observed minimum in August of 2007 (217.06 unit/l ± 8.38) and maximum in 2008 for the month of December (328.88 unit/l 6.72) and the average value of 277.26 unit/l ± 33.11 in 2007 and in 2008 the average was 289.44 unit/l ± 26.58.

At this site *Protozoa* was maximum in the month of January in 2007 (114.19 unit/l ± 0.55) and minimum was found 32.41 unit/l ± 1.23 in September month of 2007. *Rotifera* was found to be highest in the month of November 2007 (87.35 unit/l ± 1.35) and lowest (48.14 unit/l ± 1.82) in the month of September of the same year (2007). *Cladocera* was observed in their optimum peak during 2007 March (86.22 unit/l ± 1.75) and was lowest during the month of August (40.99 unit/l ± 1.47) in the year 2007. *Copepod* was maximum (55.07 unit/l ± 0.66) in 2006-2007 May and was minimum in the month of August 2007 (32.87 unit/l ± 1.53). *Ostracods sp.* was observed in their optimum peak during 2007 October (45.77 unit/l ± 1.41) and was lowest during the month of June (20.20 unit/l ± 1.27) in the year 2007.

The average value revealed that *Protozoa* was dominating in the zooplankton. In 2007 (70.56 unit/l ± 23.73) followed by *Cladocera* (67.44 unit/l ± 13.18), *Rotifera* (62.02 unit/l ± 10.00), *Copepod* (43.31 unit/l ± 7.49), *Ostracods* sp. (33.93 unit/l ± 6.64). In second year 2008 *Protozoa* (88.14 unit/l ± 13.79) found in strength followed by *Cladocera* (66.43 unit/l ± 7.13), *Rotifera* (62.72 unit/l ± 7.90), *Copepod* (44.62 unit/l ± 6.51) and *Ostracods sp* (32.77 unit/l ± 6.06).

Table 7.III.A.7: Quantitative Analysis of the Plankton of the Henwal River During 2007-2008 at sampling station III

Months	Number of Plankton per litre of water					
	Phytoplankton		Zooplankton		Total Plankton	
	2007	2008	2007	2008	2007	2008
January	923.82±2.1	945.59±2.63	306.18±3.85	305.49±5.81	1230.00±5.95	1251.08±8.44
February	968.97±1.17	1024.89±3.69	306.58±6.03	290.85±5.35	1275.45±7.2	1315.74±9.04
March	882.58±1.95	953.89±2.69	286.85±6.83	310.38±6.08	1169.43±8.78	1264.74±8.77
April	828.67±1.01	895.45±3.69	275.08±5.93	281.50±4.76	1103.75±6.94	1176.95±8.45
May	909.20±3.05	932.27±3.96	263.00±5.09	290.03±5.04	1172.20±8.14	1222.30±9.00
June	631.44±3.20	710.65±5.16	268.46±6.81	288.90±6.36	899.90±10.01	999.55±11.52
July	616.18±3.57	689.23±4.80	274.64±7.95	234.49±7.37	890.82±11.52	923.72±12.17
August	555.09±1.97	628.52±5.32	217.06±8.38	266.89±5.76	772.15±10.35	895.41±11.08
September	615.67±1.94	681.60±4.33	220.55±7.06	257.12±6.94	836.22±9.00	938.72±11.27
October	827.16±3.36	866.01±1.43	279.15±5.54	302.78±5.46	1106.31±8.90	1168.79±6.89
November	945.18±3.05	948.09±2.54	327.40±6.08	315.94±5.34	1272.67±9.13	1264.03±7.88
December	816.42±3.37	879.30±2.41	302.06±5.99	328.88±6.72	1118.48±9.36	1208.18±9.13
Average ± Sd	793.36±148.06	846.29±132.32	277.26±33.11	289.44±26.58	1070.62±175.40	1135.73±152.04

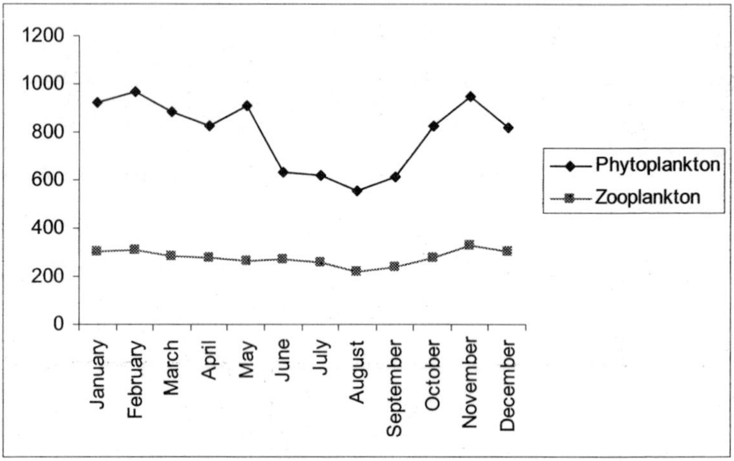

Fig. 7.III.A.13: Graph showing the monthly variation in Plankton at Sampling Site-III (2007)

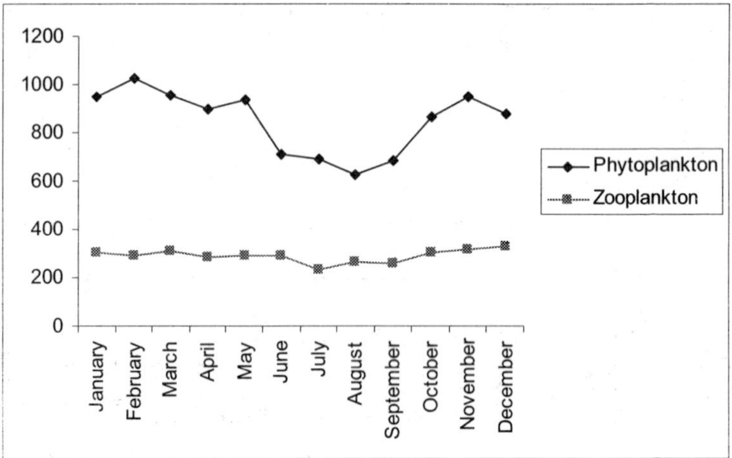

Fig. 7.III.A.14: Graph showing the monthly variation in Plankton at Sampling Site-III (2008)

Table 7.III.A.8: Number of different Group among the Phytoplankton of the Henwal River During 2007-2008 at sampling station III

Months	Total Phytoplankton/ Liter of water		Total Diatoms / Litre of water		Total green algae/ Litre of water		Total Blue green algae/ Litre of water	
	2007	2008	2007	2008	2007	2008	2007	2008
January	923.82±2.1	945.59±2.63	221.30±0.23	229.36±1.12	290.26±0.95	296.47±0.39	412.26±0.92	419.76±1.12
February	968.97±1.17	1024.89±3.69	227.68±0.23	251.42±1.06	302.41±0.69	315.72±1.58	438.78±0.25	457.75±1.05
March	882.58±1.95	953.89±2.69	239.06±0.37	290.32±0.21	285.71±0.69	292.15±1.16	357.81±0.89	371.42±1.32
April	828.67±1.01	895.45±3.69	202.29±0.27	269.02±1.13	317.25±0.29	307.78±1.25	309.13±0.45	318.65±1.31
May	909.20±3.05	932.27±3.96	204.73±0.55	242.13±1.21	291.69±1.38	290.17±1.39	412.78±1.12	399.97±1.36
June	631.44±3.20	710.65±5.16	165.52±0.37	198.35±2.45	211.21±1.61	204.76±1.20	254.71±1.22	307.54±1.51
July	616.18±3.57	689.23±4.80	132.12±0.65	162.91±2.51	187.25±1.70	190.92±1.28	296.81±1.22	335.40±1.01
August	555.09±1.97	628.52±5.32	149.24±0.19	177.37±2.83	161.53±0.71	165.70±1.24	244.32±1.07	285.45±1.25
September	615.67±1.94	681.60±4.33	177.10±0.37	201.42±1.21	181.78±0.56	187.35±1.56	256.79±1.01	292.83±1.56
October	827.16±3.36	866.01±1.43	223.89±0.19	250.97±0.29	276.45±1.36	268.92±0.15	326.82±1.81	346.12±0.99
November	945.18±3.05	948.09±2.54	262.70±0.42	277.42±0.53	305.97±1.31	301.37±0.83	376.51±1.32	369.30±1.18
December	816.42±3.37	879.30±2.41	279.30±0.27	298.75±0.36	254.63±1.92	281.91±0.43	282.49±1.18	298.64±1.62
Average±Sd	793.36±148.06	846.29±132.32	207.08±44.58	228.14±44.32	255.51±55.03	232.70±54.69	330.77±67.79	321.29±54.68

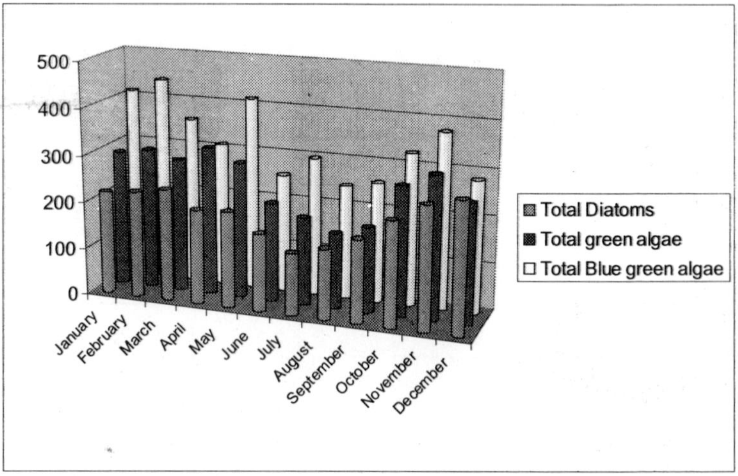

Fig. 7.III.A.15: Graph showing the monthly variation in different Groups of phytoplankton at Sampling Site-III (2007)

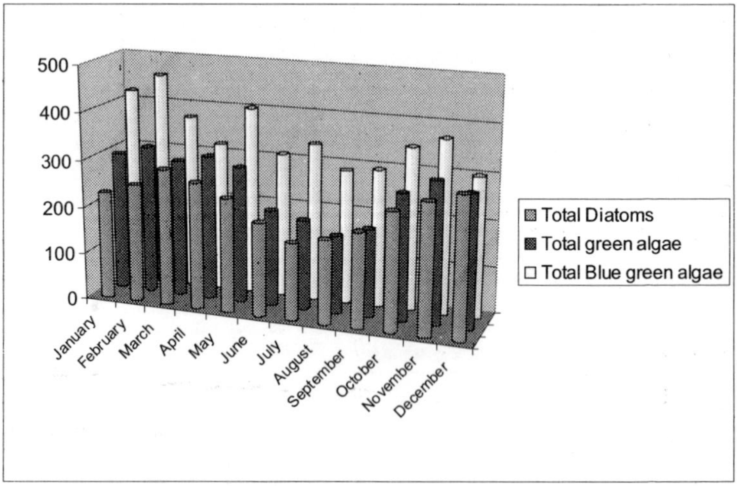

Fig. 7.III.A.16: Graph showing the monthly variation in different Groups of phytoplankton at Sampling Site-III (2008)

Table 7.III.A.9: Number of Different Genera among Zooplankton in the Henwal River During 2007-2008 at sampling station III

Months	*Protozoa*		*Rotifera*		*Cladocera*		*Copepod*		*Ostracods*	
	2007	2008	2007	2008	2007	2008	2007	2008	2007	2008
January	114.19±0.55	111.05±1.13	59.26±0.42	59.33±0.53	61.57±0.76	61.28±1.21	42.83±0.37	45.27±1.71	28.33±1.75	28.56±1.23
February	91.28±1.21	86.52±1.16	65.26±1.04	63.25±1.37	78.26±0.95	72.15±0.32	37.26±1.32	37.70±1.33	34.52±1.51	31.23±1.17
March	76.25±1.08	97.11±1.25	51.65±1.23	57.26±1.03	86.22±1.75	78.09±1.02	33.20±1.40	39.72±1.70	39.53±1.37	38.20±1.08
April	81.67±0.39	72.37±1.04	57.22±1.74	69.24±0.32	63.81±1.32	69.81±1.07	41.16±1.17	36.76±1.17	31.22±1.31	33.32±1.16
May	52.97±0.54	89.18±1.24	66.85±0.51	60.45±0.31	51.95±1.57	57.73±1.09	55.07±0.66	54.17±1.12	36.16±1.81	28.50±1.28
June	68.51±1.28	71.05±1.81	61.82±1.62	68.29±0.87	77.14±1.50	70.24±1.44	40.79±1.14	43.26±0.88	20.20±1.27	36.06±1.36
July	56.12±1.83	79.11±1.47	48.14±1.82	57.83±1.96	69.27±1.75	67.46±1.11	54.20±1.51	52.90±1.21	27.47±1.79	35.02±1.62
August	47.84±1.60	74.27±1.25	59.60±1.92	56.48±1.04	40.99±1.47	59.29±0.31	32.87±1.53	38.14±1.40	35.76±1.86	38.71±1.76
September	32.41±1.23	83.27±0.58	67.58±1.07	49.82±1.46	57.83±1.33	56.04±1.73	46.09±0.76	46.03±1.24	36.08±1.92	21.96±1.93
October	47.56±1.07	89.28±1.37	64.58±1.91	57.57±1.24	83.01±0.51	75.86±1.18	38.23±0.64	39.50±0.53	45.77±1.41	40.57±1.14
November	87.68±0.54	91.10±0.45	87.35±1.35	76.65±1.30	72.05±1.11	62.87±0.81	47.72±1.25	47.81±1.37	32.69±1.83	37.51±1.41
December	90.25±1.33	113.34±1.11	54.97±1.57	71.56±1.67	67.17±1.08	66.30±1.24	50.26±0.98	54.12±1.17	39.41±1.03	23.56±1.53
Average ± Sd	70.56±23.73	88.14±13.79	62.02±10.00	62.72±7.90	67.44±13.18	66.43±7.13	43.31±7.49	44.62±6.51	33.93±6.64	32.77±6.06

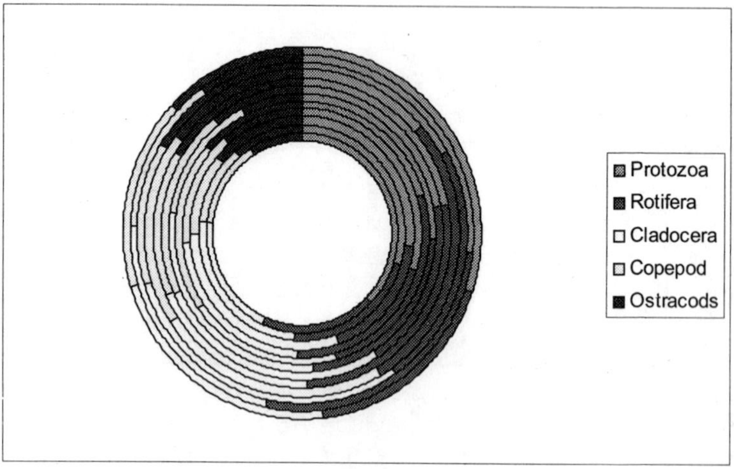

Fig. 7.III.A.17: Graph showing the monthly variation in different Genera of Zooplankton at Sampling Site-III (2007)

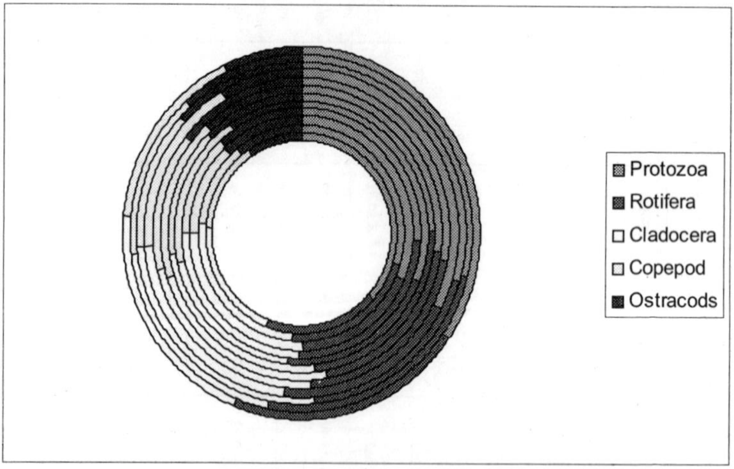

Fig. 7.III.A.18: Graph showing the monthly variation in different Genera of Zooplankton at Sampling Site-III (2008)

B. BENTHIC DIVERSITY:

Sampling site I

The comparative observation at sampling site I showed that the *Lepidoptera* shown the dominancy over the other benthic fauna with the average value of 17.00 Unit/Sq.m ± 18.84 of the year 2007 and the *Molluscs* with the minimum average value of 1.33 Unit/Sq.m ± 1.83 of the year 2007 (Table 7.III.B.1 and Figs. 7.III.B.1 to 7.III.B.4).

The *Gastropods* were observed to be nil in the months of July August and October of the year 2007 and March, April, May, June, August, September and October of 2008 and maximum 17 Unit/Sq.m. in the month of December of the year 2008. The *Molluscs* were nil in the months of May, June, July, September and October months of years 2007 and May, June and October months of 2008 and was maximum 6 Unit/Sq.m. in the months of November and December of the year 2008. The *Isopoda* were nil in the months May, June, July and August of the year 2007 and May, June, July, August and October month of 2008 and was maximum 18 Unit/Sq.m. in December of the year 2007. The *Neuroptera* found to be nil from May to October months of both the years *i.e.* 2007 and 2008 and was maximum 17 Unit/Sq.m. in the month of December of the year 2007. The *Turbellaria* were nil in the months of May, June, September and October of the year 2007 and January, May June, August-December months of 2008 and was maximum 32 Unit/Sq.m. in February of the year 2008. The *Annelida* were minimum 0 Unit/Sq.m. in the months of May, June and October of the year 2007 and March – June and August – October months of 2008 and was maximum 35 Unit/Sq.m. in February of the year 2008. The *Lepidoptera* were recorded nil in the July and August months of 2007 and April, June, July August, October and November months of 2008 and was maximum 53 Unit/Sq.m. in February of the year 2007. The total *Coleoptera* were observed nil in June and July in 2007 months of 2007 and May to September months of 2008. It was maximum in month of December with 18 Unit/Sq.m. of the year 2008.

Table 7.III.B.1: Monthly variation in Benthic Fauna of Sampling Site - I, 2007-2008

Months	*Gastropods*	*Molluscs*	*Isopoda*	*Neuroptera*	*Turbellaria*	*Annelida*	*Lepidoptera*	*Coleoptera*
2007 (Sampling Station – I)								
January	4	2	2	6	5	5	37	8
February	2	5	4	5	2	2	53	4
March	3	1	1	2	2	4	12	3
April	2	1	1	1	1	3	6	2
May	1	0	0	0	0	0	2	1
June	1	0	0	0	0	0	1	0
July	0	0	0	0	2	2	0	0
August	0	1	0	0	1	1	0	1
September	1	0	1	0	0	2	5	2
October	0	0	2	0	0	0	14	4
November	2	1	4	2	5	1	32	3
December	12	5	18	17	11	2	42	6
Average ± SD	2.33 ± 3.29	1.33± 1.83	2.75±5.01	2.75±4.94	2.42±3.23	1.83±1.59	17.00±18.84	2.83±2.41

contd. on next page...

contd...

	2008 (Sampling Station – I)							
January	8	5	4	15	0	15	15	14
February	6	4	1	6	32	35	5	6
March	0	2	6	2	15	0	1	2
April	0	4	5	7	16	0	0	1
May	0	0	0	0	0	0	2	0
June	0	0	0	0	0	0	0	0
July	2	3	0	0	8	6	0	0
August	0	1	0	0	0	0	0	0
September	0	2	2	0	0	0	2	0
October	0	0	0	0	0	0	0	2
November	6	6	1	4	0	12	0	0
December	17	6	15	15	0	24	35	18
Average± SD	3.25±5.24	2.75±2.26	2.83±4.39	4.08±5.68	5.92±10.22	7.67±11.68	5.00±10.38	3.58±6.11

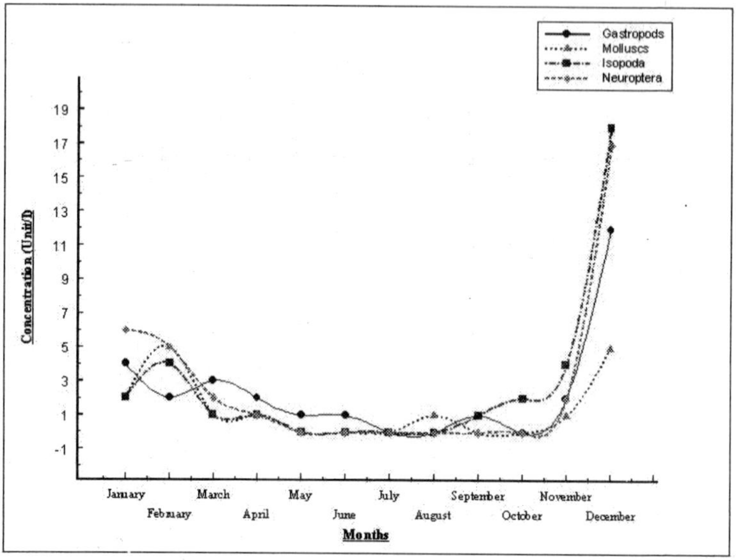

Fig. 7.III.B.1: Graph showing monthly fluctuation in *Gastropods, Molluscs, Isopoda, Neuroptera* species of Benthic Fauna at Sampling Site-I in 2007

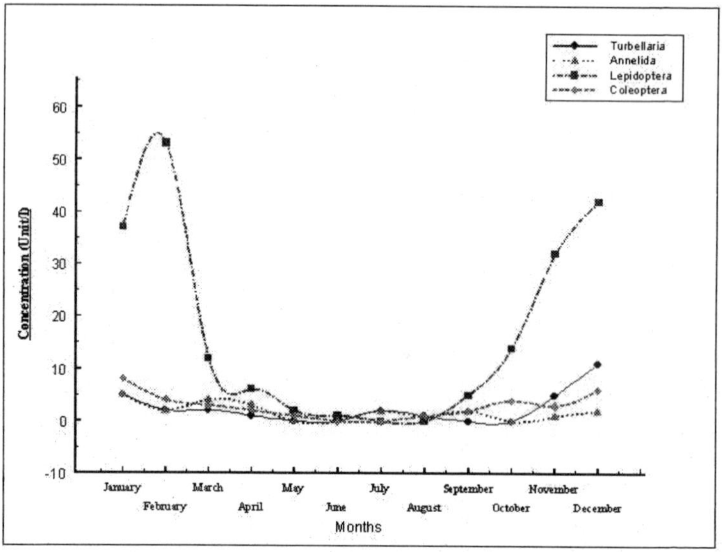

Fig. 7.III.B.2: Graph showing monthly fluctuation in *Turbellaria, Annelida, Lepidoptera, Coleoptera* species of Benthic Fauna at Sampling Site-I in 2007

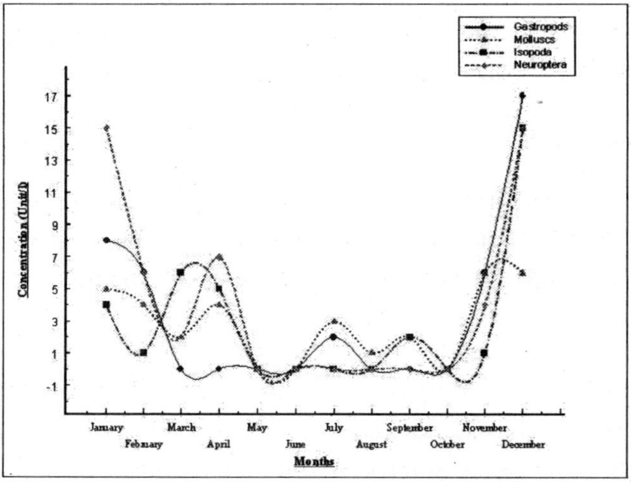

Fig. 7.III.B.3: Graph showing monthly fluctuation in *Gastropods, Molluscs, Isopoda, Neuroptera* species of Benthic Fauna at Sampling Site-I in 2008

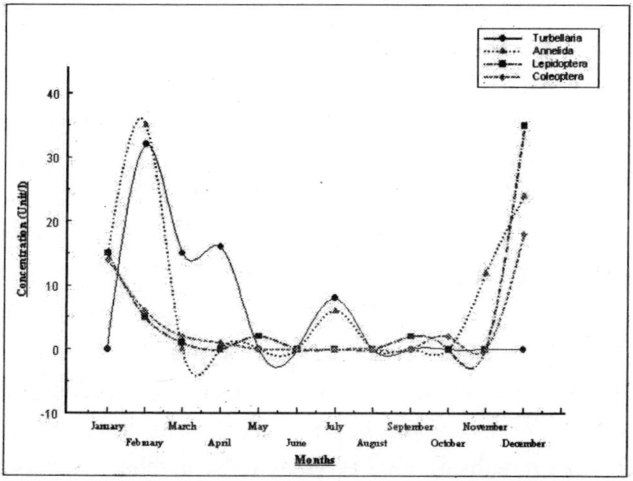

Fig. 7.III.B.4: Graph showing monthly fluctuation in *Turbellaria, Annelida, Lepidoptera, Coleoptera* species of Benthic Fauna at Sampling Site-I in 2008

Sampling site II

The comparative observation at sampling station A showed that the *Lepidoptera* shown the dominancy over the other benthicfauna with the average value of 8.58 Unit/Sq.m ± 14.51 of the year 2007 and the *Gastropods* with the minimum average value of 1.42 Unit/ Sq.m ± 1.78 of the year 2007 (Table 7.III.B.2 and Figs. 7.III.B.5 to 7.III.B.8).

The *Gastropods* were observed to be nil in the months of May to August and October month of the year 2007 and March to June and August to September months of 2008 and was maximum 24 Unit/Sq.m. in the month of February of the year 2008. The *Molluscs* were minimum nil found in the months of April, May, and June months of the year 2007 and March, April, August and October months of 2008 and was maximum 15 Unit/Sq.m. in the month of November of the year 2008. The *Isopoda* were nil in the months June, August – October months of the year 2007 and April, May, July to November months of 2008 and was maximum 24 Unit/Sq.m. in December of the year 2007. The *Neuroptera* found to be nil in the months April – June and August to October of the year 2007 and March, August, September and November months of 2008 and was maximum 30 Unit/Sq.m. in the month of December of the year 2008. The *Turbellaria* were nil in the months of January, September, October and December months of the year 2007 and June to September months of 2008 in the year 2008 and was maximum 20 Unit/Sq.m. in December month of the year 2008. The *Annelida* were nil in the months of April-June and August to November months of the year 2007 and May to October months of 2008 and was maximum 35 Unit/Sq.m. in December of the year 2008. The *Lepidoptera* were recorded nil in the September and October months of 2007 and may to September and November months of 2008 and in March of 2008 and was maximum 52 Unit/ Sq.m. in January of the year 2007. The total *Coleoptera* were observed to be nil in April, July to November months of 2007 and February, March. June, September to November months of the year 2008. It was maximum in month of December with 17 Unit/ Sq.m. of the year 2008.

Table 7.III.B.2: Monthly variation in Benthic Fauna of Sampling Station - II, 2007-2008

Months	*Gastropods*	*Molluscs*	*Isopoda*	*Neuroptera*	*Turbellaria*	*Annelida*	*Lepidoptera*	*Coleoptera*
2007 (Sampling Station – II)								
January	2	8	1	11	0	17	52	8
February	3	6	3	4	5	12	4	5
March	2	2	2	5	4	2	15	2
April	1	0	5	0	8	0	6	0
May	0	0	3	0	2	0	2	6
June	0	0	0	0	3	0	1	7
July	0	2	6	2	1	2	2	0
August	0	1	0	0	6	0	2	0
September	1	6	0	0	0	0	0	0
October	0	2	0	0	0	0	0	0
November	2	4	2	6	6	0	6	0
December	6	11	8	5	0	5	13	17
Average ± SD	1.42±1.78	3.50 ±3.56	2.5±2.65	2.75±3.52	2.92±2.84	3.17±5.61	8.58±14.51	3.75±5.21

contd. on next page...

contd...

	2008 (Sampling Station – II)							
January	8	5	6	12	6	12	10	2
February	24	4	24	6	8	8	2	0
March	0	0	3	0	5	6	6	0
April	0	0	0	5	1	2	5	2
May	0	1	0	4	16	0	0	3
June	0	1	2	6	0	0	0	0
July	2	2	0	2	0	0	0	1
August	0	0	0	0	0	0	0	6
September	0	6	0	0	0	0	0	0
October	0	0	0	1	2	0	2	0
November	16	15	0	0	1	2	0	0
December	12	12	7	30	20	35	12	5
Average± SD	5.17±7.77	3.83±4.79	3.50±6.63	5.50±8.16	4.92±6.44	5.42±9.68	3.08±4.07	1.58±2.02

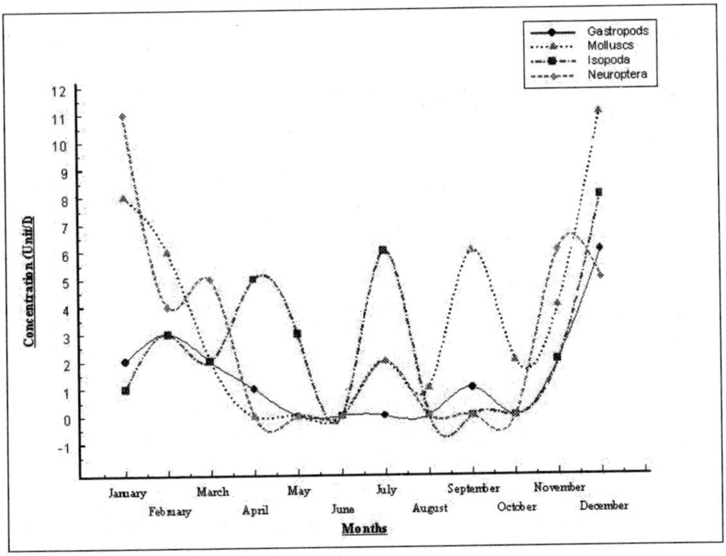

Fig. 7.III.B.5: Graph showing monthly fluctuation in *Gastropods, Molluscs, Isopoda, Neuroptera* species of Benthic Fauna at Sampling Site-II in 2007

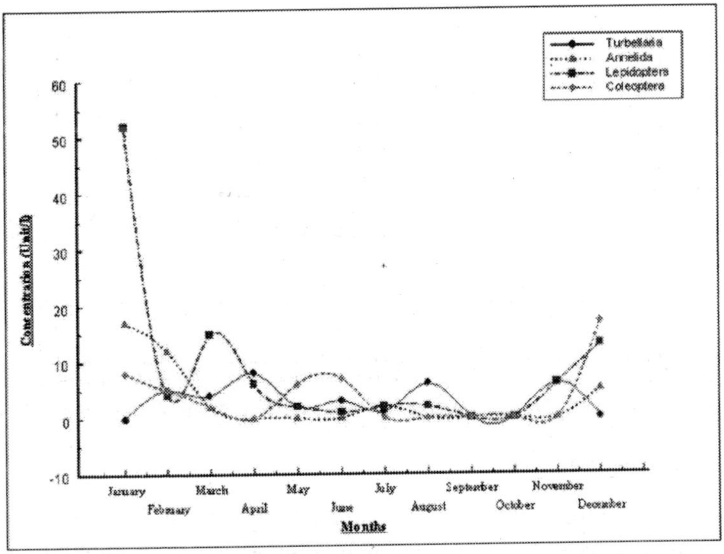

Fig. 7.III.B.6: Graph showing monthly fluctuation in *Turbellaria, Annelida, Lepidoptera, Coleoptera* species of Benthic Fauna at Sampling Site-II in 2007

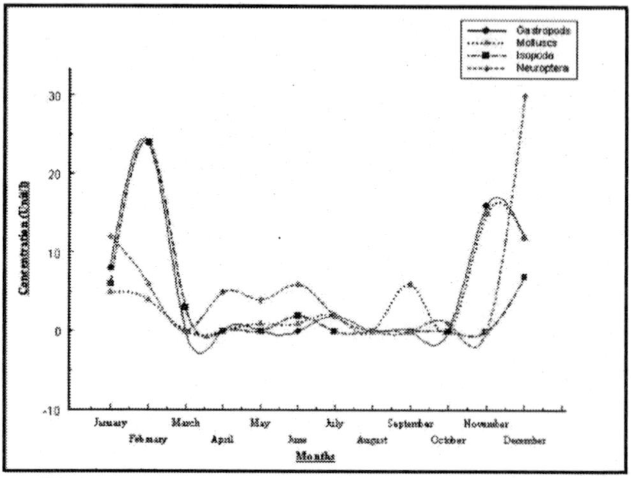

Fig. 7.III.B.7: Graph showing monthly fluctuation in *Gastropods, Molluscs, Isopoda, Neuroptera* species of Benthic Fauna at Sampling Site-II in 2008

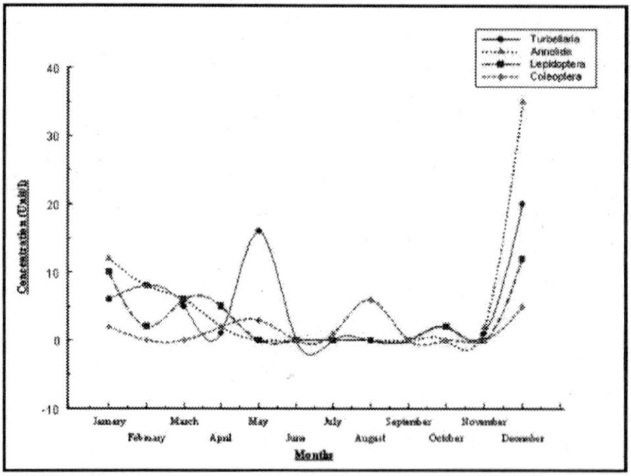

Fig. 7.III.B.8: Graph showing monthly fluctuation in *Turbellaria, Annelida, Lepidoptera, Coleoptera* species of Benthic Fauna at Sampling Site-II in 2008

Sampling site III

The comparative observation at sampling station A showed that the *Neuroptera* shown the dominancy over the other benthicfauna with the average value of 7.67 Unit/Sq.m ± 10.98 of the year 2008 and the T*urbellaria* with the minimum average value of 0.92 Unit/ Sq.m ± 1.88 of the year 2008 (Table 7.III.B.3 and Figs. 7.III.B.9 to 7.III.B.12).

The *Gastropods* were observed to be nil in the months of March to June, September and October month of the year 2007 and June to October months of 2008 and was maximum 17 Unit/ Sq.m. in the month of February of the year 2008. The *Molluscs* were minimum nil found in the months of May, June, August to October months of the year 2007 and June to November months of 2008 and was maximum 12 Unit/Sq.m. in the month of December month of the year 2008. The *Isopoda* were nil in the months May, June, August – October months of the year 2007 and April, June to September and November months of 2008 and was maximum 32 Unit/Sq.m. in December of the year 2008. The *Neuroptera* found to be nil in the months June, July September and October months of the year 2007 and May to September and November months of 2008 and was maximum 30 Unit/Sq.m. in the month of January of the year 2008. The *Turbellaria* were nil in the months of May to July and September to November months of the year 2007 and January, February, April, May and July to November months in the year 2008 and was maximum 4 Unit/ Sq.m. in January month of the year 2007. The *Annelida* were nil in the months of May, June and August to October months of the year 2007 and March to August and October months of 2008 and was maximum 12 Unit/Sq.m. in January of the year 2007. The *Lepidoptera* were recorded nil in the February to April, July and August months of 2007 and April to September and November months of 2008 and in March of 2008 and was maximum 24 Unit/ Sq.m. in January of the year 2008. The total *Coleoptera* were observed to be nil in June and August months of 2007 and May to September and November months of the year 2008. It was maximum in month of October with 14 Unit/Sq.m. of the year 2008.

Table 7.III.B.3: Monthly variation in Benthic Fauna of Sampling Station - III, 2007-2008

Months	*Gastropods*	*Molluscs*	*Isopoda*	*Neuroptera*	*Turbellaria*	*Annelida*	*Lepidoptera*	*Coleoptera*
2007 (Sampling Station – III)								
January	4	6	3	2	4	10	1	2
February	1	2	2	12	3	5	0	1
March	0	1	6	6	2	2	0	4
April	0	4	1	2	1	1	0	2
May	0	0	0	1	0	0	6	1
June	0	0	0	0	0	0	2	0
July	6	2	6	0	0	6	0	2
August	2	0	0	4	2	0	0	0
September	0	0	0	0	0	0	4	1
October	0	0	0	0	0	0	1	1
November	8	2	7	4	0	2	2	2
December	2	5	4	5	2	6	2	6
Average ± SD	1.92±2.71	1.83±2.12	2.42±2.71	3.00±3.54	1.67±1.40	2.67±3.31	1.50±1.88	1.83±1.70

contd. on next page...

contd...

	2008 (Sampling Station – III)							
January	6	2	15	30	0	12	24	6
February	17	3	6	25	0	3	1	6
March	5	4	5	16	2	0	13	10
April	2	6	0	4	0	0	0	2
May	1	2	2	0	0	0	0	0
June	0	0	0	0	6	0	0	0
July	0	0	0	0	0	0	0	0
August	0	0	0	0	0	0	0	0
September	0	0	0	0	0	4	0	0
October	0	0	2	2	0	0	1	14
November	6	0	0	0	0	2	0	0
December	5	12	32	15	3	8	5	12
Average± SD	3.50±4.95	2.42±3.60	5.17±9.52	7.67±10.98	0.92±1.88	2.42±3.90	3.67±7.45	4.17±5.29

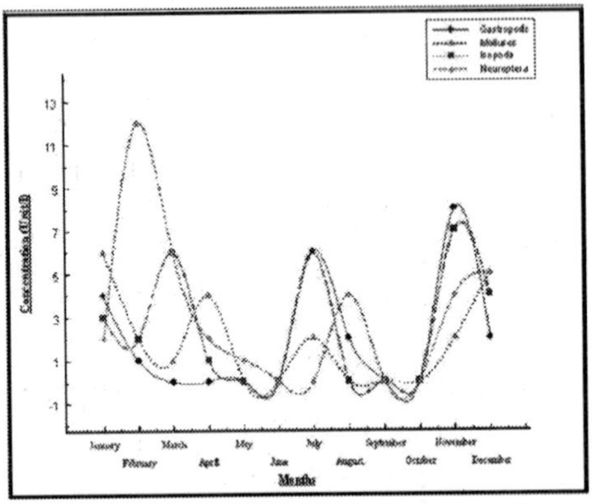

Fig. 7.III.B.9: Graph showing monthly fluctuation in *Gastropods, Molluscs, Isopoda, Neuroptera* species of Benthic Fauna at Sampling Site-III in 2007

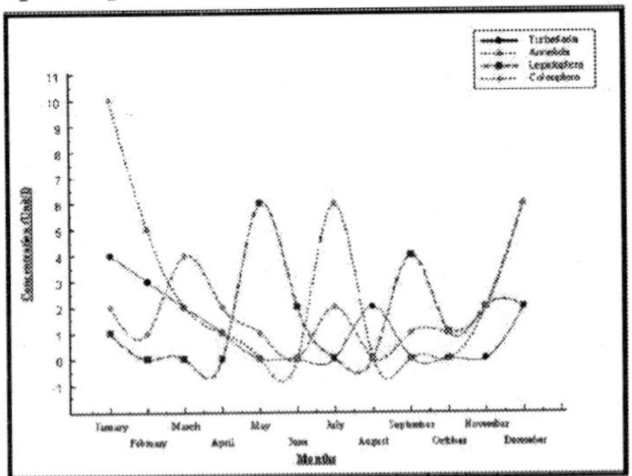

Fig. 7.III.B.10: Graph showing monthly fluctuation in *Turbellaria, Annelida, Lepidoptera, Coleoptera* species of Benthic Fauna at Sampling Site-III in 2007

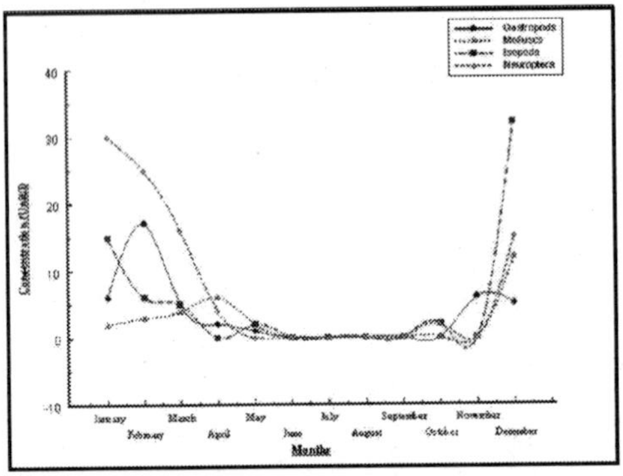

Fig. 7.III.B.11: Graph showing monthly fluctuation in *Gastropods, Molluscs, Isopoda, Neuroptera* species of Benthic Fauna at Sampling Site-III in 2008

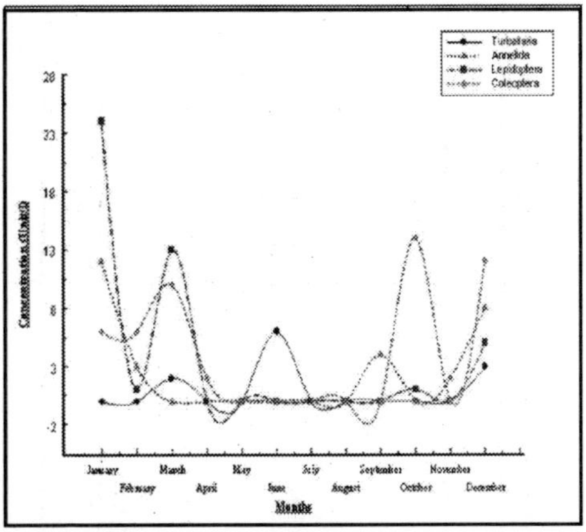

Fig. 7.III.B.12: Graph showing monthly fluctuation in *Turbellaria, Annelida, Lepidoptera, Coleoptera* species of Benthic Fauna at Sampling Site-III in 2008

C. FISH DIVERSITY:

During the present study period different species of fish fauna were collected from the different sites of Henwal River. Fishes were collected from various spots of River Henwal, all through the sampling range. The river contained rich fish diversity and population. A systematic list of fishes found in River Henwal was arranged and given below in Table 7.III.C.1.

Table 7.III.C.1. Systematic list of fishes found in the river Henwal

Phylum	:	Vertebrata
Sub-phylum	:	Craniata
Super-class	:	Gnathostomata
Series	:	Piscean
Class	:	Telcostomi
Sub-class	:	Actinopterygiie
Order	:	Cypriniformes
Sub-order	:	Cyprinoidei
Family	:	Cyprinidae
Sub-family	:	Raborinae
Genus	:	*Barilius Hamilton*

Present scientific name	**Name according to Day**	**Name according to Hamilton**	**Local name**
Barilius bendelensis (Ham.)	*Barilius bendelisis*	*Cyprinus bendelisis*	Chal
Raimas bola (Ham.) (Endangered)	*Barilius bola*	*Cyprinus bola*	Bhola
Barilius vagra (Ham.)	*Barilius vagra*	**Cyprinus vagra**	Chal
-	*Barilius modestus*		
Barilius barna	-	-	Pota
	Genus	*Danio Hamilton*	
Danio (Danio) devario (Ham.)	*Damino devario*	*Cyprinius devario*	Patukari
	Sub-family	Cypriniae	
	Genus	*Chaganio smith*	
Chagunio chagunio (Ham)	*Barbus chagunio*	*Cyprinius chagunio*	Gelhari
	Genus	*Labeo Cuvier*	
Labeo dero (Ham.)	*L. dero*	*C. dero*	
Labeo dyocheilus (Mc. Clell)	*L. dyocheilus*	*L. dyocheilus*	

contd. on next page...

Eromus danricus (Ham.)	-	-	
	Genus	*Crossocheilus Vanhasselt*	
Crossocheilus latius latius (Ham.)	*Cirrhinus latia*	*Cirrhinus latius*	Pet phomi
-	-	*C. gohan*	
-	-	*C. sada*	
-	*Cirrhinus mrigala*	*C. mrigala*	Nain
	Genus	***Gara Hamilton***	
Garra gotyla gotyla (Gray)	*Discognathus lamata*	*C. lamata*	Gunthale
	Genus	*Tor gray*	
Tor putitora (Endangered)	*Barbus tor*	*C. tor*	Mahaseer
Tor tor (Ham.) (Endangered)	*Barbus tor*	*C. hexasticus*	Mahaseer
Puntius sarna (Ham.)	*B. sarna*	*C. sarna*	Darahi
Puntius chola (Ham.)	*B. chola*	*C. chola*	Shidhari
Puntius conchonus (Ham.)			Darahi

contd. on next page...

Puntius ticto (Ham.)	*B. ticto*	*C. ticto*	Sidhari
	Sub-family	**Schyzothoracinae heckel**	
	Genus	***Schizothorax heckel***	
Schyzothorax plagiostomus	*Oreinus plagiostomus*		
Schyzothorax sinatus	*Oreinus sinatus*		
	Family	**Cobitidae**	
	Genus	***Botia gray***	
Botia dario (Ham.)	*Botia dario*	*Cabitus geto*	Baghaua
Lepdocephalus guntea (Ham.)	*L. guntea*	*L. guntea*	Hakati
	Family	Schilbeidae	
	Genus	*Clupisoma swainson*	
Clupisoma garua (Ham.)	*Pseudeutropius garva*	*Gilurus garua*	Raikar karahi
	Order	Perciformes	
	Sub-order	Anabantoidei	

contd. on next page...

	Family	Anabanlidae	
	Genus	*Colisa cuver etvalenciennes*	
Closia fasciatuas (Block & Schn.)	*Trichogaster fasciatus*	*Trichopodius colisa*	Khosti
-	-	*T.bejeus*	
-	-	*T. corta*	
	Order	**Mastacembeliformes**	
	Family	**Mastacembelidas**	
	Genus	***Mastacembelus schopoli***	
Mastacembelus armatus (Lacepede)	*Mastacembelus armatus*	*Macrognathus armatus*	Bam
Mastacembelus punctatus (Ham.)	*Mastacembelus punctatus*	*Macrognathus punctatus*	Malga
	Family	Bragridae	
Mystus seenghala (Sykes)	*Macrones seenghala*	-	Tengar
Mystus bleekeri (Day)	*Macrones bleekeri*	-	Tengar
Mystus tengara (Ham.)	*Macrones tengara*	-	Tengar
Mystus vittatus (Bl.)	*Macrones vittatus*	-	Tengar
Mystus aor (Ham.)	*Macrones aor*	-	Tengar

CONCLUSION

The information given in the review of literature represents the studies done on the riverine system and the biota in it. The present study is devoted for evaluation of different physical, chemical and biological parameters of the River Henwal.

Among all factors controlling the environment of water reservoir, temperature is one of the major factors having the most intense direct and indirect effect on the water reservoir on any aquatic ecosystem. In fact, no other single factor has so intense influence and direct and indirect effect on the biota of an ecosystem. In the present study of River Henwal, a difference in the fluctuation of water temperature was observed 13.0°C ± 0.25 (minimum) in the month of December month of 2008 at study site – I and 29.0°C ± 0.50 (maximum) in June at study site - III. Annual average values of temperature varied between 20.25°C ± 4.38 to 20.83°C ± 4.23. Minimum annual average value of temperature observed in the year 2008 at sampling site-I and maximum was observed in the same year in 2008 at sampling site-III. The water temperature showed an upward trend from January to April followed by a downward from May onwards. A more or less similar trend has been observed in the river Yamuna by Chakrabarty *et al.* (1959) and in the Kallayi (John, 1976). Badola and Singh (1981) also reported similar trend in the river Alaknanda. Same study was made by Singh *et al.* (1988; 1989) in river Ganga, Yamuna and Sangam and Gautam *et al.* (2000) in Ganga River at Rishikesh. Same trend of temperature was observed by Khanna *et al.* (2001) in river Ganga at Haridwar.

The electrical conductivity has always been used as a valuable method to estimate the degree of salt contents and total dissolved solids in water. During the study of River Henwal the conductivity fluctuation was observed 96.45 Siemens/cm ±0.05 to 131.57 Siemens/cm ± 0.05. The maximum value was recorded in October month of 2007 at study site – I and minimum value was recorded at study site – III in February month of 2007. It can be said that the present higher conductivity values in August to October month due to the input of large amount of salts and silt carried by the river. Identical results were observed by Das *et al.* (1992), Bhattacharya and Saha (1988), Raina *et al.* (1982) and Mittal and Sengar (1990) from various Indian rivers. Khanna *et al.* (2003) also observed similar trend of conductivity in Ganga River at Bulandshahar. The annual average value conductivity in River Henwal is ranging between 114.92 Siemens/cm ± 9.51 to 116.25 Siemens/cm ± 10.30.

The water of the River Henwal becomes start turbid from June month onward and in July to September the water was highly turbid. It showed that turbidity and total solids were closely inter related with one another and cause common effect upon the river and aquatic life as also stated by Verma *et al.* (1984). In the current study of river Henwal the minimum value of turbidity was recorded 16.79 JTU ± 1.91 in the month of February of year 2007 at study site – II. The highest value was found (376.46 JTU ± 178.06) in the month of August at study site – II in 2007. Annual average value of turbidity varied between 149.94 JTU ± 112.84 to 172.13 JTU ± 116.99 in which maximum average value observed at sampling site - III in the year 2008 and minimum at sampling site – I in year 2007. Similar pattern was also reported by Badola and Singh (1981), Dobriyal *et al.* (1983) in the hill streams of the Garhwal Himalaya, Khanna *et al.* (1997) in river Ganga, and Seth *et al.* (2000) in the river Ganga. The turbidity is a striking characteristic to know the physical status of the rivers. The suspended particles, soil particles, discharged effluents; decomposed organic matter, total dissolved solids as well as the microscopic organisms increase the turbidity of water, which interferes with the penetration of light. The increased turbidity during monsoon season was attributed due to heavy soil erosion in the nearby catchments and massive contribution of suspended solids from sewage canals. Surface runoff and domestic wastes mainly contribute to the increased turbidity of the River Henwal. Bhatt *et al.* (1984) observed low turbidity during November, February and an increase after May and reaches peak in August in river Kosi. He attributed that during monsoon months, the water contained large amount of silt, fine sand particles, organic matter and clay. In river Ganga and its tributaries, minimum values for turbidity were recorded in winter and summer season while maximum in monsoon period (Bilgrami and Duttamunshi, 1985). Higher values were recorded at places where human activity was intense or at places where effluents from industries/farms are discharged into the river. High amount of silt carried due to fast currents of Kosi attributed mainly for the turbidity of this river (Bhatt *et al.*, 1984).

During the course of study of River Henwal the minimum value of water velocity was recorded in the February month of year 2007 with 0.56 m/sec ± 0.35 at study site –I and the maximum water velocity was noted in September month of 2007 with 1.20 m/sec ± 0.04 at study site – II. The annual average of water velocity during the study area of river Henwal was ranging between 0.80 m/sec ± 0.20 to 0.88 m/sec ± 0.19. The minimum annual average value of velocity was recorded in 2007 at study site – I and maximum annual average at study site – II in 2008. The velocity started increasing from May onwards due to melting of snow at the place of origin of the river Henwal. Again beyond September, the velocity started decreasing. In the present study the velocity and turbidity showed a direct relationship. Similar observations have been recorded by Dobriyal (1985) and Nautiyal (1990) in the snow fed rivers of the Garhwal Himalaya.

In the present investigation it was noted that the total solids were maximum in the month of July (534.71 mg/l ± 2.40) of year 2008 at study site - I, which were responsible for the turbidity in the river Henwal. The total solids were recorded minimum (51.76 mg/l ± 0.78) in the month of February of year 2008 at study site – I due to gradual sedimentation of the filterable residue. The annual average values of total solids for the current study period are varied between 234.34 mg/l ± 150.53 to 249.78 mg/l ± 170.56. Minimum average value of total solids observed in the year 2007 at study site – II and maximum in the year 2008 at study site - III. Khanna and Singh (2000) found that total solids were maximum in summer in the water of Suswa River. Total solids cause ecological imbalance in the aquatic ecosystem by mechanical abrasive action. Higher values of total solids may cause deterioration of the surviving condition of aquatic organisms by mechanical abrasive action and enhance the turbidity of the river. Same conditions were shown by Khanna (1993), Verma and Shukla (1969) in their studies. Zingde *et al.* (1980), Kudesia and Verma (1985) and Reddy and Venkateswarlu (1987) studied that most of the Indian rivers show similar tendency with respect to fluctuations of total solids. Khanna *et al.* (1997) and Seth *et al.* (2000) also made out the same study. Similar trends were shown by Khanna and Chugh (2004) during study of water quality of River Ganga at Haridwar.

The utilization of dissolved oxygen is in the respiration of the plants and animals and also in the oxidation process of break down which results in lowering of concentration of dissolved oxygen in the water. Temperature is also closely related to oxygen.

In warm water with adequate oxygen the break down process occurs rapidly. An inverse correlation was observed throughout the study period between temperature and dissolved oxygen. Dissolve oxygen is also an important parameter which speaks health of the water body in terms of trophic status, and also in biotic status (Edmondson 1961, Nalewazko *et al.,* 1981).

During the course of study the maximum value of dissolved oxygen was recorded (10.4 mg/l ± 0.17) in the month of December at study site – I in the year 2008. The minimum value of dissolved oxygen (7.1 mg/l ± 0.16) was observed in the month of April of the same year and at the same site *i.e.* sampling site - I. The annual average value of dissolved oxygen ranged between 8.73 mg/l ± 0.75 to 9.04 mg/l ± 0.90. The minimum annual average value of DO was observed in the year 2007 at sampling site – III and maximum was observed in 2008 at sampling site - II. The dissolved oxygen reduced gradually from May onward due to turbidity which retarded the photosynthetic activity of aquatic flora. The temperature showed an inverse relationship with the DO almost throughout the study. The cause of maximum dissolved oxygen in November to February is due to reduced rate of decomposition by decreased microbial activity at low temperature (Strommer and Smock, 1989). Chopra *et al.* (1990), Gopal and Sah (1993) and Sharma (1999) also have got the same result and have opined that low temperature in November to February increases the oxygen retaining capacity of water and solubility of O_2 in water. This trend was also observed by Badola and Singh (1981) in the river Alaknanda. Khanna (1993) has also reported the same trends in the river Ganga at Haridwar. Same study is also made by Pandey *et al.* (2003) in Ganga canal at Hardwar. Concentration of dissolved oxygen is one of the most important parameter to indicate water purity and to determine the distribution and abundance of various algal groups. It has been recommended that a minimum of 4 mg/l of dissolved oxygen should be maintained in water for healthy growth of fish and other microbial population. Different workers have pointed out various influencing factors on oxygen level that include discharges from industries, water current, velocity and biota.

Ray *et al.* (1966) have reported that the Ganga River contains maximum free carbon dioxide (free CO_2) in the rainy season in Allahabad. Chakrabarty *et al.* (1959) also recorded the maximum free carbon dioxide in Yamuna during monsoon season at Allahabad. The free carbon dioxide is released during the decomposition of certain substances and because of respiration of the prevailing living organisms. Since higher temperature regulates the decomposition of organic substances and also the respiration of the living biota, a direct relationship existed between free carbon dioxide and water temperature. The maximum free carbon dioxide was observed (1.9 mg/l ± 0.89) in May month in 2008 at sampling site - III i.e. summer season because of the slightly higher temperature of water and higher turbidity, which lowers the rate of photosynthesis and thus the dissolved oxygen. However, it was minimum (0.1 mg/l) in the January month of sampling site – III in both years and in January at sampling site - I. The free CO_2 thus showed an inverse relationship with dissolved oxygen. The similar trend was observed by Khanna (1993). Dorris (1963) reported that increase in turbidity due to washing of organic matter that is brought along with rain, would reduce the rate of photosynthesis due to less penetration of sun light which results in less oxygen and more carbon dioxide in water.

Many workers have also reported that this rise in free carbon dioxide is toxic for fishes. Powers (1938) reported cases of sudden mortality of fishes, which might have happened due to high levels of free carbon dioxide in water and quenching of dissolved oxygen in water. Hynes (1970) declared, dissolved oxygen and free carbon dioxide are usually inversely proportional to one another. The reason for the fluctuation in dissolved oxygen may be due to total organic carbon load, the bottom plankton and benthic fauna etc. Chandler (1940) observed plankton in relation to some physico-chemical parameters and supported the above observation.

The annual average value of free carbon dioxide varied between 0.50 mg/l ± 0.57 at sampling site –III in 2007 to 0.80 mg/l ± 0.75 in which maximum average value obtained from sampling site – I in the year 2007 in the River Henwal. Pahwa and Mehrotra (1966) and Ray *et al.* (1966) have reported that the Ganga River contains maximum free carbon dioxide in rainy season at Allahabad. Khanna *et al.* (1997) and Seth *et al.* (2000) have also reported the same trends in the river Ganga at Haridwar but Khanna and Singh (2000) observed maximum free carbon dioxide during summer in Suswa River at Raiwala, Dehradun. The free carbon dioxide is released during the decomposition of certain substances and metabolic activity of the living organisms. Since higher temperature accelerate the decomposition of organic substances as well as the respiratory activity of the biota, a direct relationship was established between the water temperatures and free carbon dioxide. The dissolved oxygen and free carbon dioxide usually inversely related to one another because of photosynthetic and respiratory activity of the biota (Hynes, 1970).

Many studies also reveal that the alkalinity may also increase due to decomposition of organic matter that has settled at the bottom. The total alkalinity was maximum (229.00 mg/l ± 3.17) in the month of May in the year 2007 at sampling site – III and was minimum with 109.00 mg/l ± 0.12) in September month of year 2007 at sampling site - I. In the period of lower rate of precipitation accompanied with warmer and humid climatic condition, the degradation of organic matter is enhanced. Whereas during the winter season the organic matter at the bottom is difficult to degrade because of low temperature of water. The annual average value of alkalinity varied between 173.67 mg/l ± 36.63 to 181.42 mg/l ± 30.29 in which minimum average value is obtained from sampling site - I in the year 2007 and maximum from sampling site – III in the year 2008. Similar observations were also obtained by Holden and Green (1960), Talling and Rzoska (1967) and Abdin (1948 a,b). The decomposition of the organic matter leads to the high alkalinity of the waters. (Venkateswarlu & Jayanti, 1968).

In the current limnological study of river Henwal the maximum values of total hardness were recorded in May (256.00 mg/l ± 0.16) of year 2008 at sampling site – III and the minimum value recorded was in March (108.00 mg/l ± 0.00) in the year 2007 at sampling site - III. The annual average values of total hardness ranged from 167.59 mg/l ± 37.27 to 177.00 mg/l ± 42.94 in which maximum average value is recorded from sampling site-III in 2008 and the minimum value was recorded at sampling site-I in the same year. Calcium and Magnesium are the two component of total hardness. It is present in form of carbonates and bicarbonates. Calcium is one of the most abundant substances of natural waters. Being present in high quantities in the rocks, it is leached from there to contaminate the water. Calcium is essential for metabolic processes in all-living organisms. In this study the Calcium found in range of 19.1 mg/l ± 0.30 in March of year 2008 at sampling site – I to 46.7 mg/l ± 2.30 in February at sampling site – I in the year 2007. Maximum values of Magnesium were recorded in June (28.37 mg/l ±0.25) of the year 2007 at sampling site – III and the minimum value was observed in month of March (11.89 mg/l ± 0.24) at sampling site – I in the year 2007. Khanna *et al.* (1993, 2003) and Mishra and Joshi (2003) observed hardness in river Ganga at Haridwar and found more or less similar trends in their study. The annual average value of calcium hardness is ranged between 31.42 mg/l ± 7.02 to 33.81 mg/l ± 6.24. The maximum value was observed at the sampling site of the study in the year 2008 and minimum at sampling site – II in the same year. The annual average of magnesium hardness was fluctuating between 19.20 mg/l ± 4.53 to 20.88 mg/l ± 3.95. The maximum annual average was recorded in the second year of the study *i.e.* at sampling site – II and the minimum value was recorded in the sampling site –I of the year 2007.

The highest and lowest values of acidity were found in February and November month of years 2007 and 2008 at the same sampling site *i.e.* sampling site – I with 60.32 mg/l ± 0.00 and 27.54 mg/l ± 0.06 respectively. The annual average value of acidity varied between 42.33 mg/l ± 8.82 to 45.47 mg/l ± 10.22 in which maximum average value was recorded from the sampling site - I in the year 2007 and minimum was at sampling site - II in the year 2008.

The maximum value of chloride was found in the month of October (13.21 mg/l ± 3.51) and minimum in July (5.01 mg/l ± 0.21) at sampling site – I in the years 2008 and 2007 respectively. The annual average value of chloride varied between 7.82 mg/l ± 0.94 to 8.22 mg/l ± 1.81 in which maximum average value was recorded from the sampling site – I in the year 2008 and minimum was at sampling site – III in the year 2007. Chlorides are present in sewage, sewage effluents and farm drainage. Significant levels of chloride were shown by many rivers like Yamuna (Sengar *et al.*, 1885); Tungbhadra (Reddy and Venkateswarlu, 1987); Jhelum (Raina *et al.*, 1984) and Kshipra (Mishra and Saxena, 1984). CPCB (2003) reported the value of chloride in between 14 to 51 mg/l during Ganga monitoring from Bithur, Kanpur to Sangam, Allahabad. CPCB (1990-91) also studied the chloride from Rishikesh to Hardwar and showed a significant increasing trend on chloride at all monitoring stations in West Bengal stretches.

The phytoplankton of many Indian rivers consists of diverse assemblage of major taxonomic groups. Many of these forms have different environmental and physiological requirements. The number, type and distribution of these organisms present in any aquatic habitat provide a clue on the environmental conditions prevailing in that particular habitat. It is seen that many environmental factors interact to provide conditions for the growth of plankton both specially and seasonally. Khanna (1993) emphasized the need for monitoring of a river continuously at least for ten years to assess the importance of various chemical parameters in the growth and distribution of plankton. Ray (1955) and Venkateswarlu and Menon (1969) recorded maximum values of total plankton during winter and minimum during rainy season.

In the present work it has been noted that the maximum total phytoplankton were 1067.12 Unit/l ±1.99 were found in the month of December at sampling site – I in the year 2007 and zooplankton were 342.93 Unit/l ± 5.01 in the month of December at sampling site – I in the year 2008. The annual average value of phytoplankton ranged between 777.61 Unit/l ± 204.67 to 846.29 Unit/l ± 132.32 in which minimum average value was recorded from sampling site- II in the year 2007 and maximum was recorded from sampling site - III in the year 2008. The annual average value of zooplankton ranged between 275.91 Unit/l ± 35.56 to 294.09 Unit/l ± 27.16 in which maximum average value was obtained from sampling site – I in 2007 and minimum was obtained from sampling site in the same year 2007. The study showed a dominance of phytoplankton as also reported by Pahwa and Mehrotra (1966) in the Ganga River. Badola and Singh (1981), Dobriyal and Singh (1987), Khanna (1993) and Khanna *et al.* (2000) observed high values of plankton during January-March. Khanna and Singh (2000) also reported high values of plankton during winter season.

The Planktonic concentration also affected by the turbidity. The increased turbidity reduces the plankton production (Khanna *et al.*, 1993). In the present study it was noted that temperature show a negative relationship with plankton. The plankton was maximum in November to March when the temperature was low. Eddy (1934) and Chandler (1940) pointed out that the plankton production is mainly influenced by temperature. Khanna (1993), Khanna *et al.* (2000) also reported the same trends. The results indicated that the plankton were maximum in the winter month probably due to low temperature, high content of DO and low velocity. Similar study was made by Khanna and Bhutiani (2003) and Khanna *et al.* (2000).

The benthic fauna consists of minute organisms that move around in water. These constitute the main bulk of food of bottom feeding fishes and are relatively less exposed to environmental fluctuations than the organisms on the surface water, thus they serve as bioindicators of ecological conditions of a habitat. Singh *et al.* (1994) also performed a qualitative and quantitative study of benthic fauna of Kararia Lake. As per Singh and Singh (2002) these are of great importance for biological monitoring. These serve as food for fish community, along with plankton. Thus they serve as an important section of water ecology. In Henwal river water, the production of benthic fauna is mainly influenced by temperature and light. Jindal and Singh (2005) had similar observations in a study of fresh water ecosystem. Temperature and sunlight have also a major contribution in productivity and growth of the alga, hydrophytes and other phytoplankton. Although the temperature of river is low in winters even then due to increased water transparency the solar rays and energy can be transmitted more, which is a reason for the flourishing phytoplankton and hydrophytes in winters (Khanna, 1993). These may serve as support for the life of benthic fauna, providing food, shelter and also required DO which increases in water due to higher rate of photosynthesis in Winter season. Thus in the river Henwal, the availability of food for the icthyofauna is almost continuous in all the seasons.

In the present study the benthic fauna followed the trend in increasing order from *Lepidoptera* > *Neuroptera* > *Annelida* > *Gastropods* > *Isopoda* > *Molluscs* > *Coleoptera* > *Turbellaria* .

This can be due to increased velocity of flooded river which swipes away most of the phytoplankton, macro-invertebrates, nymphs, insect larvae and diatoms. This finding is similar to that of Khanna (1993). Prasad and Singh (2003) observed that as the effect of increased volume and velocity of water plankton and benthic fauna were disturbed and get decreased. In some places Annelids and Gastropods were observed during Monsoon. Higher suspended solids serve as a blanket, reduces the penetration of solar light and energy required for growth of hydrophytes and phytoplankton, this interferes with photosynthesis activity and also reduces the oxygen concentration in water therefore the Benthic fauna gets reduced. In winter season, the situation is however, vice versa as because of low turbidity sunrays can penetrate deeper in water and increase photosynthetic activity. This was also reported by Bisht and Das (1985), Grey and Ward (1982) and Roy_(1996). During the study period *Ephimeroptera (*May fly nymphs) seemed to dominate. Kumar and Bohra (1999) represented Gastropoda as indicators of pollution status of some wet lands in Santhal Pargana. They said that Gastropoda (*Melania striatella*) are rare and tolerant to heavy organic pollution. Jindal and Singh (2005) showed *Oligochaetes* and insects positive corelation with smaller soil particles while *Molluscs* with larger soil particles. Various other workers like Sharma *et al.* (2002) observed the biodiversity of benthic communities of Sirpur Lake. Singh *et al.* (2002) have also performed study on distribution and seasonal abundance of benthic macro invertebrates in the Bisar sarovar and made attempts to find the effect of aquatic pollution on benthic fauna.

The flood is also harmful for the fishes as reported by Verma and Dalela (1975 a) the suspended, fine or coarse solids in floodwater may create mechanical injuries to the gills of the fishes. Badola (1979) reported that high turbidity and flash floods might even kill the fishes as mentioned in the incident of the Alaknanda River during 1970. Because of the washing away and disturbance of the phytoplankton and other macro-invertebrates the chances of scarcity of fish food is higher in Monsoon months. This may be a limiting factor for the growth of fishes in addition to the other anthropogenic activities, which disturb and destruct the fish ecology. Sivakumar (2003) made observations on some biological aspects of the fresh water fish. During the present study period different species of fish fauna were collected from the different sites of Henwal River. Fishes were collected from various spots of River Henwal, all through the sampling range. The river contained rich fish diversity and population.

Table 1: Annual average of Physical Parameters for different sampling sites during 2006-2008

Parameters	*Sampling Site – I*	*Sampling Site – II*	*Sampling Site – III*
	2007		
Water velocity (m/sec)	0.80±0.20	0.83±0.20	0.84±0.19
Water Temperature (°C)	20.46±4.62	20.66±4.12	20.33±5.08
Conductivity (Siemens/cm)	115.74±11.66	114.92±9.51	116.04±22.29
Turbidity (JTU)	149.94±112.84	159.98±123.19	159.01±104.17
Total solids (mg/l)	236.79±169.65	234.34±150.53	245.55±163.41
	2008		
Water velocity (m/sec)	0.86±0.17	0.88±0.19	0.88±0.16
Water Temperature (°C)	20.25±4.38	20.50±5.12	20.83±4.23
Conductivity (Siemens/cm)	115.41±11.70	115.15±11.40	116.25±10.30
Turbidity (JTU)	154.00±112.91	165.79±120.65	172.13±116.99
Total solids (mg/l)	244.44±177.58	247.65±167.36	249.78±170.56

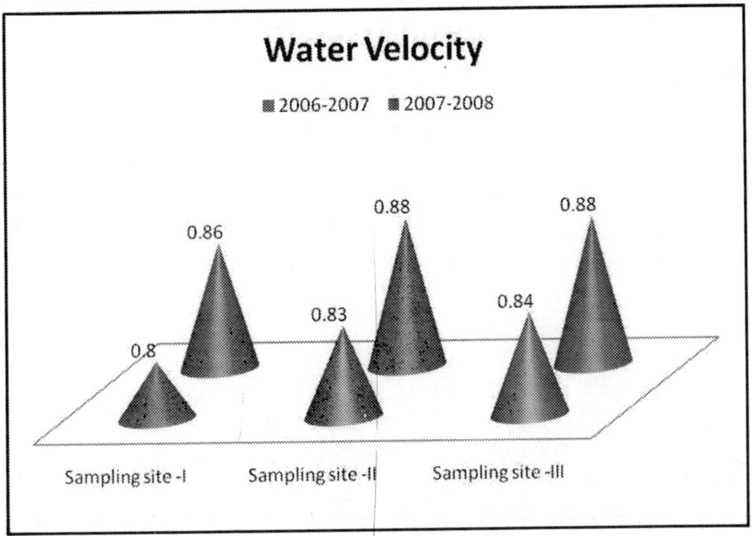

Fig. 1 Average value of Water Velocity of Henwal River at different sampling sites during 2007-2008

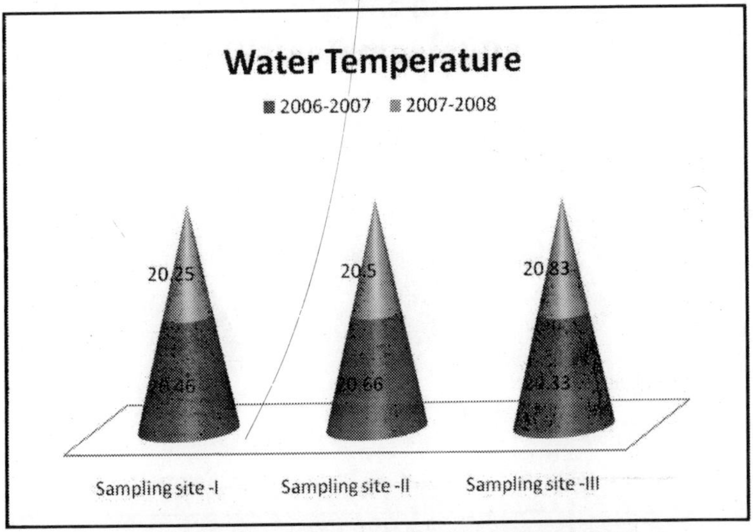

Fig. 2 Average value of Water Temperature of Henwal River at different sampling sites during 2007-2008

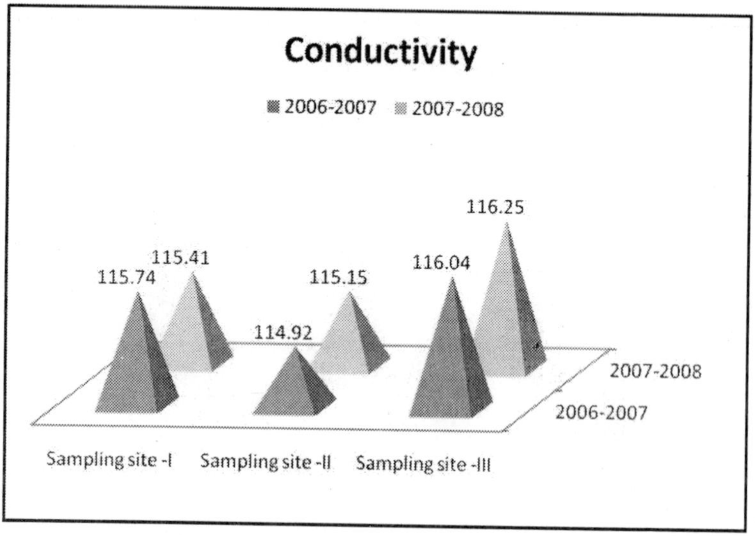

Fig. 3 Average value of Conductivity of Henwal River at different sampling sites during 2007-2008

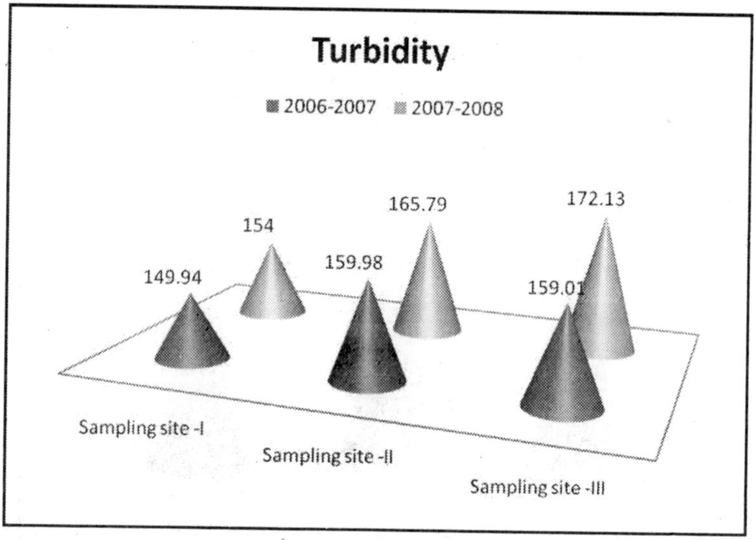

Fig. 4 Average value of Turbidity of Henwal River at different sampling sites during 2007-2008

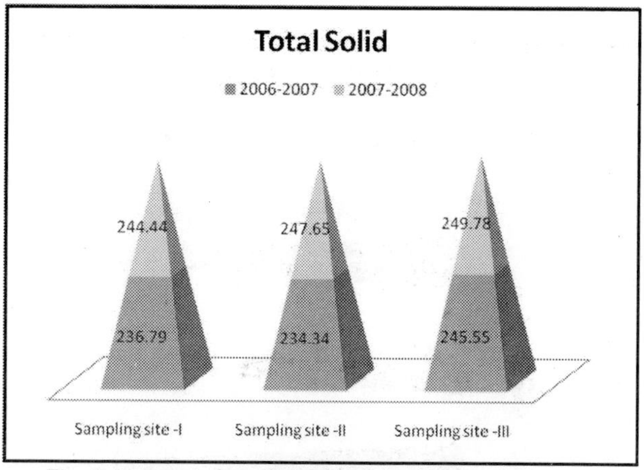

Fig. 5 Average value of Total Solids of Henwal River at different sampling sites during 2007-2008

Table 2: Annual average of Chemical Parameters for different sampling sites during 2007-2008

Parameters	Sampling Site –I	Sampling Site –II	Sampling Site –III
	2007		
DO (mg/l)	8.83±0.86	8.98±0.87	8.73±0.75
Free CO_2(mg/l)	0.48±0.75	0.45±0.73	0.34±0.57
Acidity (mg/l)	45.47±10.22	43.81±8.29	45.17±9.02
Total Alkalinity (mg/l)	173.67±36.63	175.31±30.84	181.29±36.78
Total Hardness (mg/l)	170.46±33.51	169.83±30.37	171.17±40.96
Ca Hardness (mg/l)	31.59±8.03	33.45±6.25	32.08±5.33
Mg Hardness (mg/l)	19.20±4.53	19.91±4.52	20.33±4.67
Chlorides (mg/l)	7.92±1.87	7.88±1.29	7.82±0.94
Phosphates (mg/l)	0.52±0.20	0.53±0.20	0.54±0.17
	2008		
DO (mg/l)	8.78±0.92	9.04±0.90	8.75±0.73
Free CO_2(mg/l)	0.44±0.73	0.44±0.72	0.36±0.61
Acidity (mg/l)	44.18±9.11	42.33±8.82	44.68±7.65
Total Alkalinity (mg/l)	176.25±26.34	178.92±27.83	181.42±30.29
Total Hardness (mg/l)	167.59±37.27	170.50±29.75	177.00±42.94
Ca Hardness (mg/l)	31.63±8.28	31.42±7.02	33.81±6.24
Mg Hardness (mg/l)	19.58±4.43	20.88±3.95	20.64±4.16
Chlorides (mg/l)	8.22±1.81	7.94±1.72	7.92±1.05
Phosphates (mg/l)	0.51±0.18	0.55±0.19	0.52±0.15

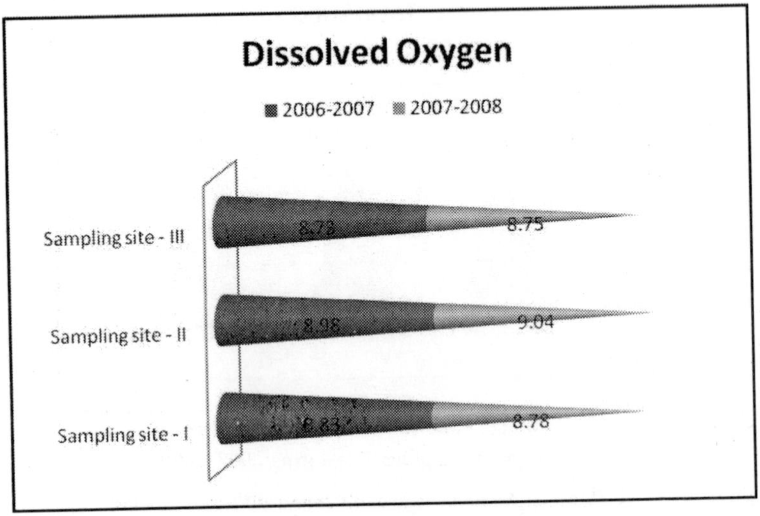

Fig. 6 Average value of Dissolved Oxygen of Henwal River at different sampling sites during 2007-2008

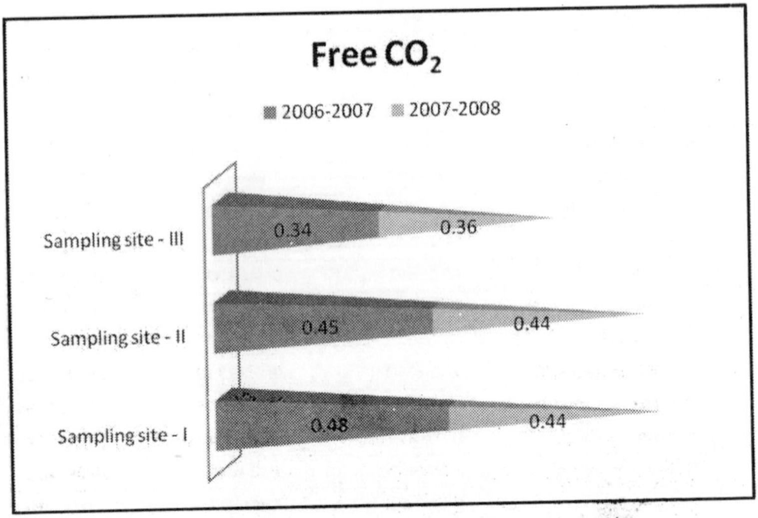

Fig. 7 Average value of Free CO_2 of Henwal River at different sampling sites during 2007-2008

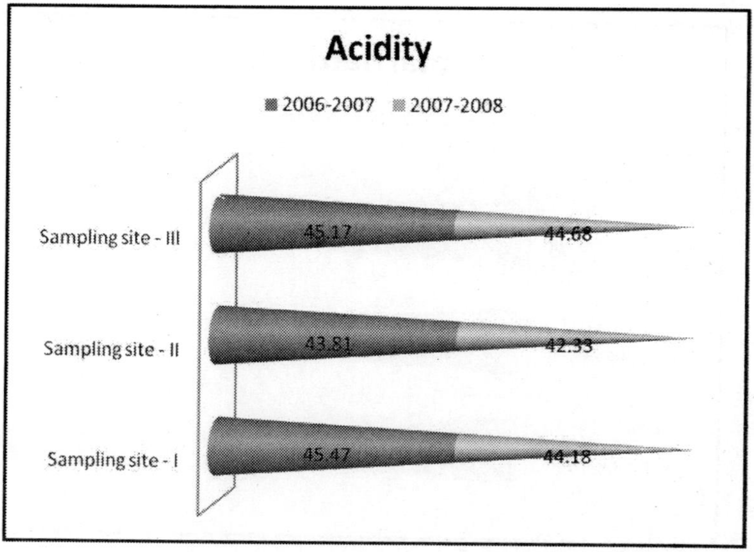

Fig. 8 Average value of Acidity of Henwal River at different sampling sites during 2007-2008

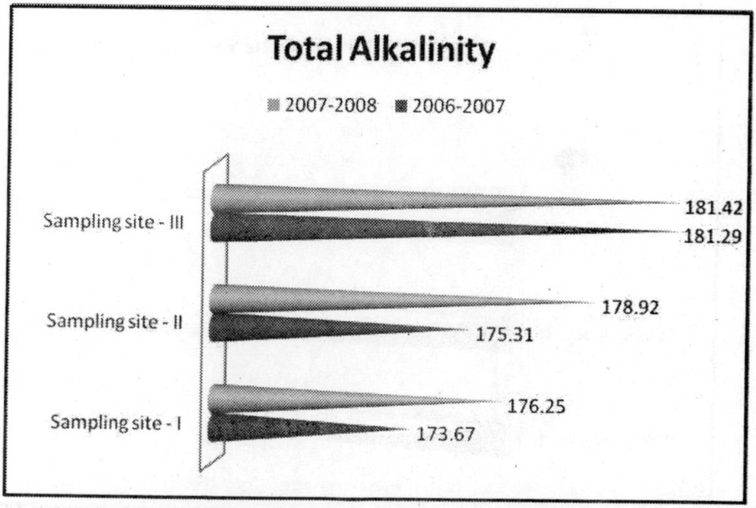

Fig. 9 Average value of Total Alkalinity of Henwal River at different sampling sites during 2007-2008

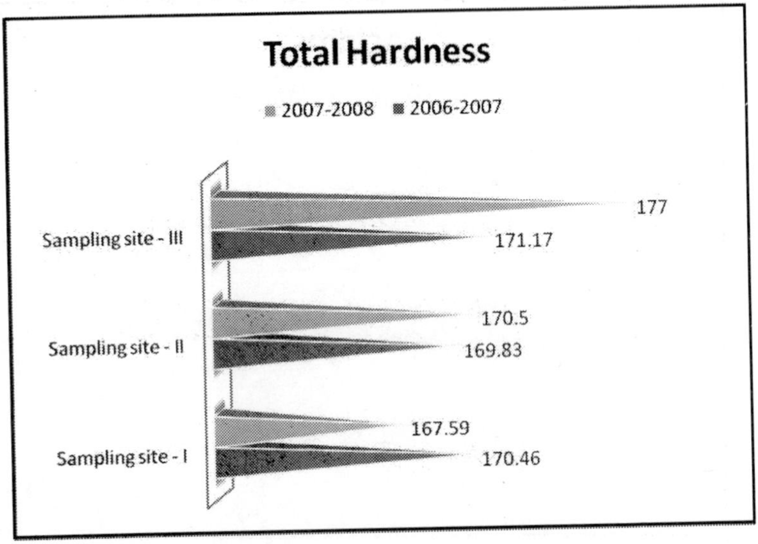

Fig. 10 Average value of Total Hardness of Henwal River at different sampling sites during 2007-2008

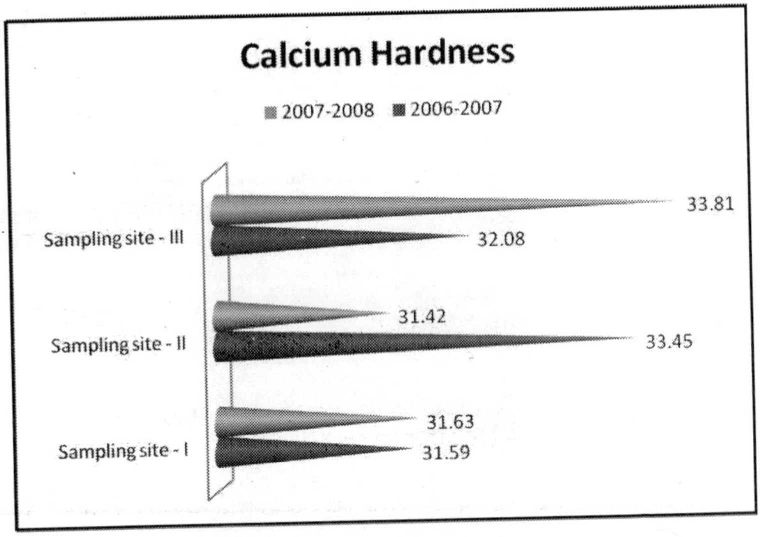

Fig. 11 Average value of Calcium Hardness of Henwal River at different sampling sites during 2007-2008

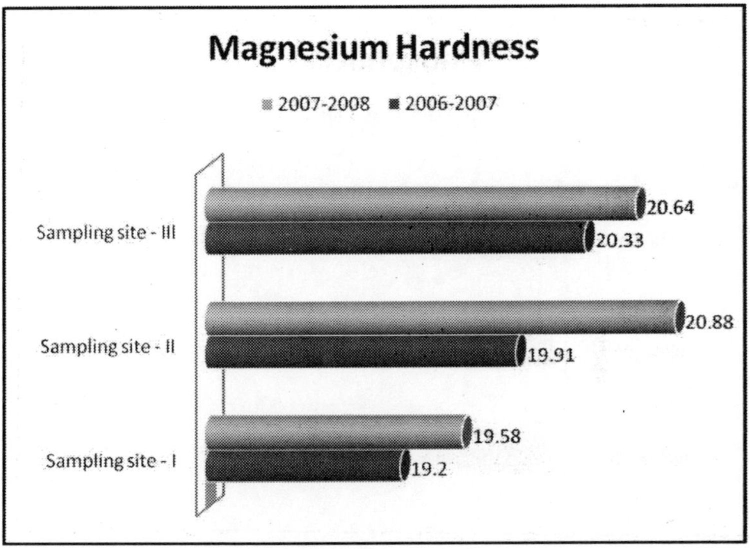

Fig. 12 Average value of Magnesium Hardness of Henwal River at different sampling sites during 2007-2008

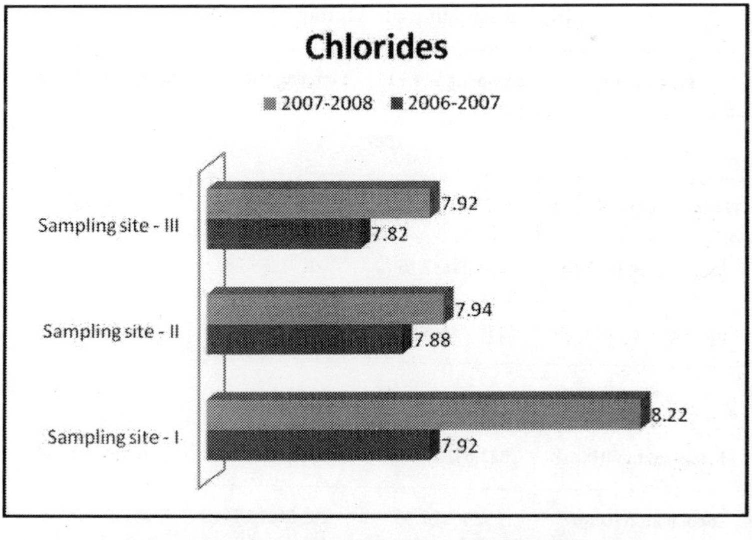

Fig. 13 Average value of Chlorides of Henwal River at different sampling sites during 2007-2008

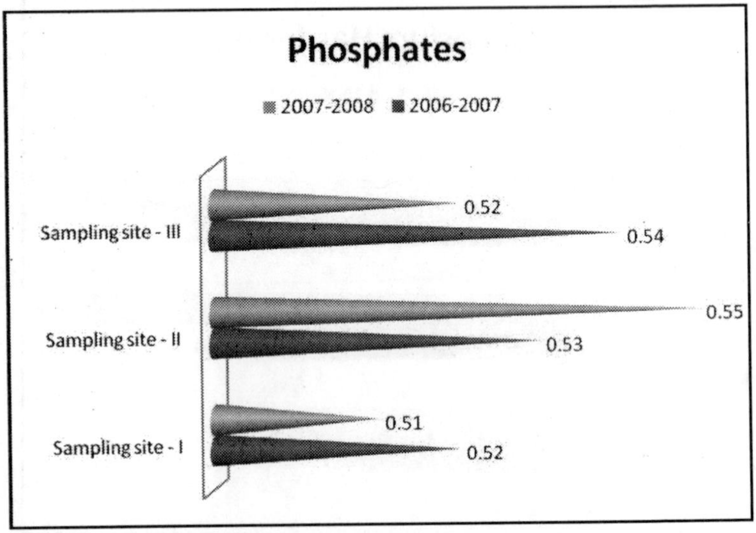

Fig. 14 Average value of Phosphates of Henwal River at different sampling sites during 2007-2008

Table 3: Annual average of Plankton for different sampling sites during 2007-2008

Parameters	Sampling Site – I	Sampling Site – II	Sampling Site – III
2007			
Phytoplankton (Unit/l)	822.14±163.54	777.61±204.67	793.36±148.06
Zooplankton (Unit/l)	294.09±27.16	275.91±35.56	277.26±33.11
Total Plankton (Unit/l)	1116.23±179.25	1053.52±225.84	1070.62±175.40
2008			
Phytoplankton (Unit/l)	823.69±167.26	789.13±222.19	846.29±132.32
Zooplankton (Unit/l)	292.22±27.83	289.56±28.78	289.44±26.58
Total Plankton (Unit/l)	1115.92±189.08	1078.69±234.17	1135.73±152.04

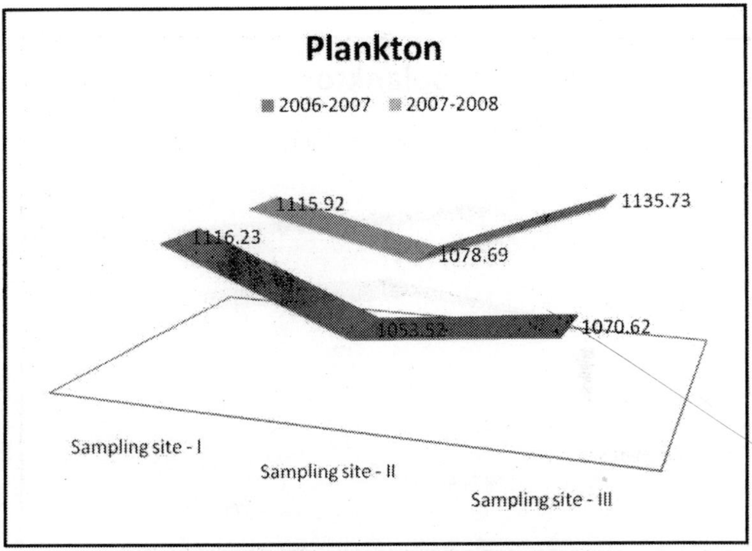

Fig. 15 Average value of Plankton of Henwal River at different sampling sites during 2007-2008

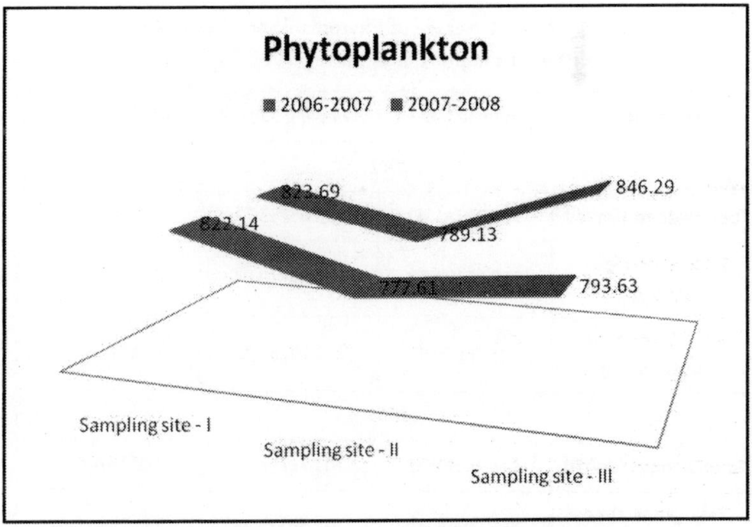

Fig. 16 Average value of Phytoplankton of Henwal River at different sampling sites during 2007-2008

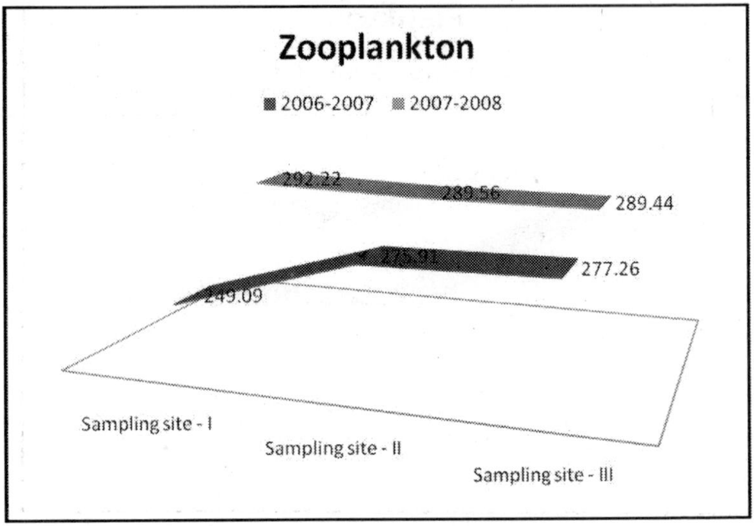

Fig. 17 Average value of Zooplankton of Henwal River at different sampling sites during 2007-2008

Table 4 : Annual average of Phytoplankton Groups for different sampling sites during 2007-2008

Parameters	Sampling Site – I	Sampling Site – II	Sampling Site – III
2007			
Total Diatoms (Unit/l)	231.39±44.00	201.65±53.73	207.08±44.58
Total green algae (Unit/l)	257.05±72.79	252.76±79.84	255.51±55.03
Total Blue green algae (Unit/l)	333.70±76.99	320.69±97.35	330.77±67.79
2008			
Total Diatoms (Unit/l)	226.25±40.02	203.13±66.08	228.14±44.32
Total green algae (Unit/l)	265.02±76.42	264.49±78.85	232.70±54.69
Total Blue green algae (Unit/l)	332.42±85.58	268.57±98.37	321.29±54.68

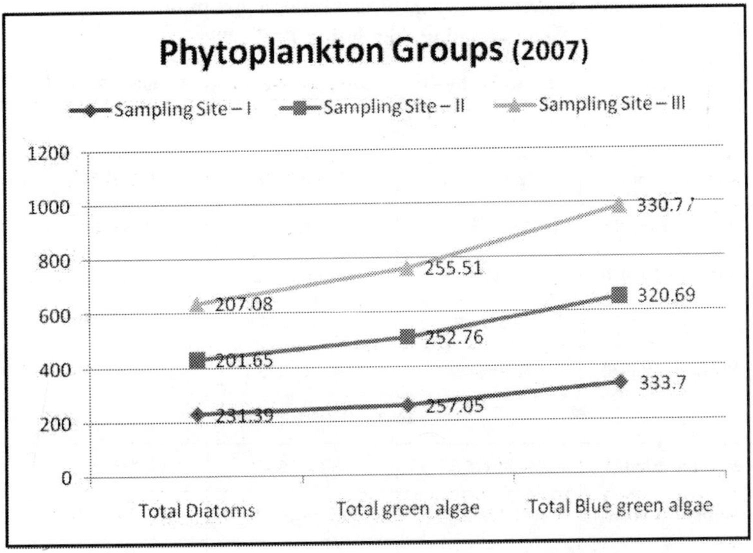

Fig. 18 Average value of Phytoplankton Groups of Henwal River at different sampling sites during 2007

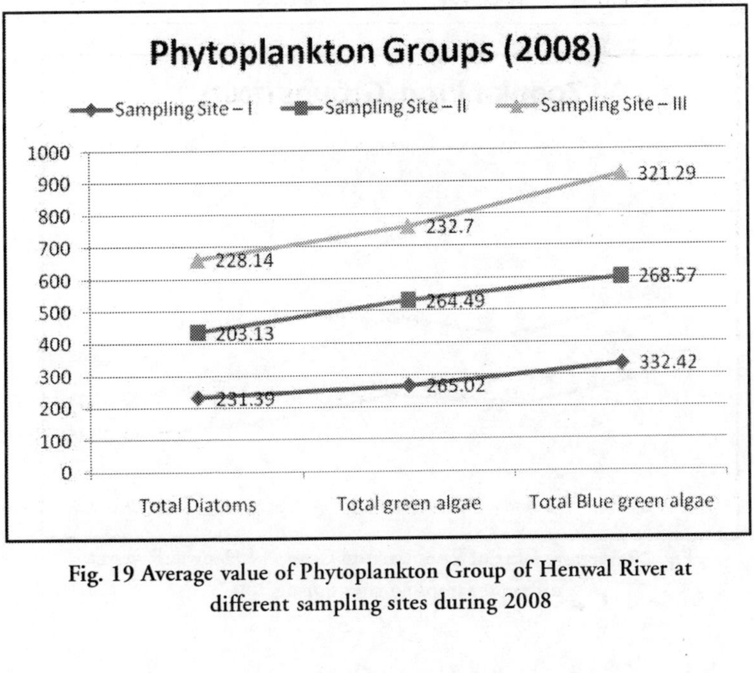

Fig. 19 Average value of Phytoplankton Group of Henwal River at different sampling sites during 2008

Table 5 : Annual average of zooplankton groups for different sampling sites during 2007-2008

Parameters	Sampling Site – I	Sampling Site – II	Sampling Site – III
2007			
***Protozoa* (Unit/l)**	87.13±14.62	71.95±22.58	70.56±23.73
***Rotifera* (Unit/l)**	61.67±8.65	59.54±9.57	62.02±10.00
***Cladocera* (Unit/l)**	68.45±10.83	68.36±11.30	67.44±13.18
***Copepod* (Unit/l)**	44.39±7.94	44.56±8.24	43.31±7.49
***Ostracods* (Unit/l)**	32.45±4.90	31.51±7.35	33.93±6.64
2008			
***Protozoa* (Unit/l)**	86.49±10.93	88.64±16.32	88.14±13.79
***Rotifera* (Unit/l)**	61.12±8.10	60.64±9.11	62.72±7.90
***Cladocera* (Unit/l)**	66.54±9.21	65.26±10.25	66.43±7.13
***Copepod* (Unit/l)**	43.65±8.01	41.53±8.43	44.62±6.51
***Ostracods* (Unit/l)**	34.43±4.99	33.48±6.13	32.77±6.06

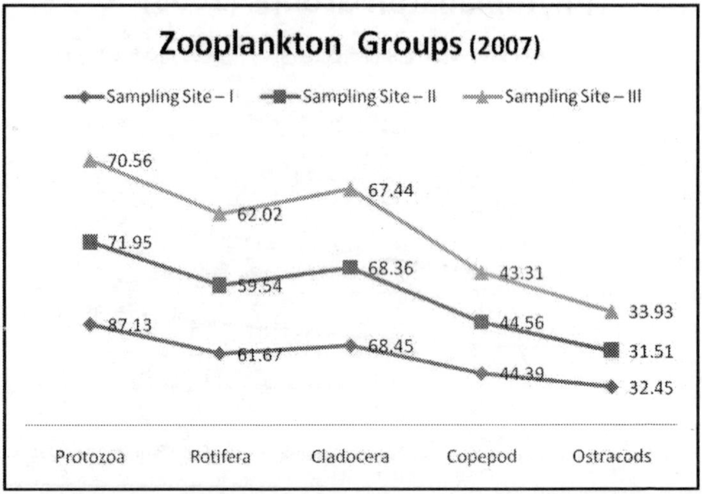

Fig. 20 Average value of Zooplankton Group of Henwal River at different sampling sites during 2007

Fig. 21 Average value of Zooplankton group of Henwal River at different sampling sites during 2008

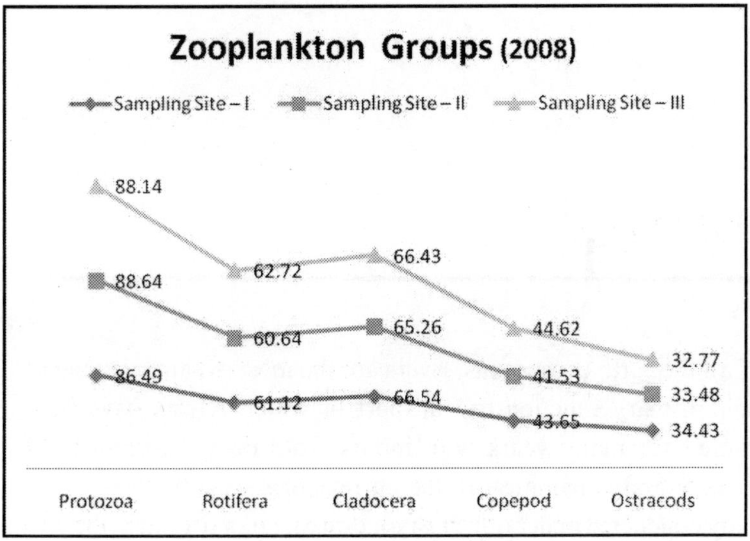

SUMMARY

Of all the earth ecosystems, rivers are the most dynamic having as their primary function to transport of water. Rivers have been studied for many years with inputs from many disciplines of sciences and humanities. In the current thesis the findings of the study conducted under title "**LIMNOLOGICAL CHARACTERIZATION AND ITS RELATION TO BIODIVERSITY OF SPRING FED RIVER THE HENWAL A TRIBUTARY TO RIVER GANGA OF GARHWAL HIMALAYA**" during January 2007 to December 2008. In this study several aspects of the ecological factors which affect the water system through direct or indirect ways, are analyzed and evaluated in various aspects.

The results which were observed during this study are summarized as below:

- In the present study of River Henwal, a difference in the fluctuation of water temperature was observed 13.0°C ± 0.25 (minimum) in the month of December month of 2008 at study site - I and 29.0°C ± 0.50 (maximum) in June at study site - III. Annual average values of temperature varied between 20.25°C ± 4.38 to 20.83°C ± 4.23. Minimum annual average value of temperature observed in the year 2008 at sampling site -I and maximum was observed in the same year in 2008 at sampling site - III.

- During the study of River Henwal the conductivity fluctuation was observed 96.45 Siemens/cm ±0.05 to 131.57 Siemens/cm ± 0.05. The maximum value was recorded in October month of 2007 at study site – I and minimum value was recorded at study site – III in February month of 2007. It can be said that the present higher conductivity values in August to October month due to the input of large amount of salts and silt carried by the river. The annual average value conductivity in River Henwal is ranging between 114.92 Siemens/cm ± 9.51 to 116.25 Siemens/cm ± 10.30.
- In the current study of river Henwal the minimum value of turbidity was recorded 16.79 JTU ± 1.91 in the month of February of year 2007 at study site – II. The highest value was found (376.46 JTU ± 178.06) in the month of August at study site – II in 2007. Annual average value of turbidity varied between 149.94 JTU ± 112.84 to 172.13 JTU ± 116.99 in which maximum average value observed at sampling site - III in the year 2008 and minimum at sampling site – I in year 2007.
- During the course of study of River Henwal the minimum value of water velocity was recorded in the February month of year 2007 with 0.56 m/sec ± 0.35 at study site –I and the maximum water velocity was noted in September month of 2007 with 1.20 m/sec ± 0.04 at study site – II. The annual average of water velocity during the study area of river Henwal was ranging between 0.80 m/sec ± 0.20 to 0.88 m/sec ± 0.19. The minimum annual average value of velocity was recorded in 2007 at study site – I and maximum annual average at study site – II in 2008. The velocity started increasing from May onwards due to melting of snow at the place of origin of the river Henwal. Again beyond September, the velocity started decreasing.
- In the present investigation it was noted that the total solids were maximum in the month of July (534.71 mg/l ± 2.40)

of year 2008 at study site - I, which were responsible for the turbidity in the river Henwal. The total solids were recorded minimum (51.76 mg/l ± 0.78) in the month of February of year 2008 at study site – I due to gradual sedimentation of the filterable residue. The annual average values of total solids for the current study period are varied between 234.34 mg/l ± 150.53 to 249.78 mg/l ± 170.56. Minimum average value of total solids observed in the year 2007 at study site – II and maximum in the year 2008 at study site - III.

- During the course of study the maximum value of dissolved oxygen was recorded (10.4 mg/l ± 0.17) in the month of December at study site – I in the year 2008. The minimum value of dissolved oxygen (7.1 mg/l ± 0.16) was observed in the month of April of the same year and at the same site *i.e.* sampling site - I. The annual average value of dissolved oxygen ranged between 8.73 mg/l ± 0.75 to 9.04 mg/l ± 0.90. The minimum annual average value of DO was observed in the year 2007 at sampling site – III and maximum was observed in 2008 at sampling site - II.
- The maximum free carbon dioxide was observed (1.9 mg/l ± 0.89) in May month in 2008 at sampling site - III i.e. summer season because of the slightly higher temperature of water and higher turbidity, which lowers the rate of photosynthesis and thus the dissolved oxygen. However, it was minimum (0.1 mg/l) in the January month of sampling site – III in both years and in January at sampling site - I. The free CO_2 thus showed an inverse relationship with dissolved oxygen. The annual average value of free carbon dioxide varied between 0.50 mg/l ± 0.57 at sampling site –III in 2007 to 0.80 mg/l ± 0.75 in which maximum average value obtained from sampling site – I in the year 2007 in the River Henwal.
- Many studies also reveal that the alkalinity may also increase due to decomposition of organic matter that has

settled at the bottom. The total alkalinity was maximum (229.00 mg/l ± 3.17) in the month of May in the year 2007 at sampling site – III and was minimum with 109.00 mg/l ± 0.12) in September month of year 2007 at sampling site - I. In the period of lower rate of precipitation accompanied with warmer and humid climatic condition, the degradation of organic matter is enhanced. The annual average value of alkalinity varied between 173.67 mg/l ± 36.63 to 181.42 mg/l ± 30.29 in which minimum average value is obtained from sampling site - I in the year 2007 and maximum from sampling site – III in the year 2008.

- In the current limnological study of river Henwal the maximum values of total hardness were recorded in May (256.00 mg/l ± 0.16) of year 2008 at sampling site – III and the minimum value recorded was in March (108.00 mg/l ± 0.00) in the year 2007 at sampling site - III. The annual average values of total hardness ranged from 167.59 mg/l ± 37.27 to 177.00 mg/l ± 42.94 in which maximum average value is recorded from sampling site - III in 2008 and the minimum value was recorded at sampling site - I in the same year. Calcium and Magnesium are the two component of total hardness. In this study the Calcium found in range of 19.1 mg/l ± 0.30 in March of year 2008 at sampling site – I to 46.7 mg/l ± 2.30 in February at sampling site – I in the year 2007. Maximum values of Magnesium were recorded in June (28.37 mg/l ±0.25) of the year 2007 at sampling site – III and the minimum value was observed in month of March (11.89 mg/l ± 0.24) at sampling site – I in the year 2007.
- The annual average value of calcium hardness is ranged between 31.42 mg/l ± 7.02 to 33.81 mg/l ± 6.24. The maximum value was observed at the sampling site of the study in the year 2008 and minimum at sampling site – II in the same year. The annual average of magnesium hardness was fluctuating between 19.20 mg/l ± 4.53 to

20.88 mg/l ± 3.95. The maximum annual average was recorded in the second year of the study *i.e.* at sampling site – II and the minimum value was recorded in the sampling site –I of the year 2007.

- The highest and lowest values of acidity were found in February and November month of years 2007 and 2008 at the same sampling site *i.e.* sampling site – I with 60.32 mg/l ± 0.00 and 27.54 mg/l ± 0.06 respectively. The annual average value of acidity varied between 42.33 mg/l ± 8.82 to 45.47 mg/l ± 10.22 in which maximum average value was recorded from the sampling site - I in the year 2007 and minimum was at sampling site - II in the year 2008.
- The maximum value of chloride was found in the month of October (13.21 mg/l ± 3.51) and minimum in July (5.01 mg/l ± 0.21) at sampling site – I in the years 2008 and 2007 respectively. The annual average value of chloride varied between 7.82 mg/l ± 0.94 to 8.22 mg/l ± 1.81 in which maximum average value was recorded from the sampling site – I in the year 2008 and minimum was at sampling site – III in the year 2007.
- In the present work it has been noted that the maximum total phytoplankton were 1067.12 Unit/l ±1.99 were found in the month of December at sampling site – I in the year 2007 and zooplankton were 342.93 Unit/l ± 5.01 in the month of December at sampling site – I in the year 2008. The annual average value of phytoplankton ranged between 777.61 Unit/l ± 204.67 to 846.29 Unit/l ± 132.32 in which minimum average value was recorded from sampling site-II in the year 2007 and maximum was recorded from sampling site - III in the year 2008. The annual average value of zooplankton ranged between 275.91 Unit/l ± 35.56 to 294.09 Unit/l ± 27.16 in which maximum average value was obtained from sampling site – I in 2007 and minimum was obtained from sampling site in the same year 2007.

- In the present study the benthic fauna followed the trend in increasing order from *Lepidoptera* > *Neuroptera* > *Annelida* > *Gastropods* > *Isopoda* > *Molluscs* > *Coleoptera* > *Turbellaria*. This can be due to increased velocity of flooded river which swipes away most of the phytoplankton, macro-invertebrates, nymphs, insect larvae and diatoms. This finding is similar to that of Khanna (1993). Prasad and Singh (2003) observed that as the effect of increased volume and velocity of water plankton and benthic fauna were disturbed and get decreased.
- The flood is also harmful for the fishes as reported by Verma and Dalela (1975 a) the suspended, fine or coarse solids in floodwater may create mechanical injuries to the gills of the fishes. Badola (1979) reported that high turbidity and flash floods might even kill the fishes as mentioned in the incident of the Alaknanda River during 1970. Because of the washing away and disturbance of the phytoplankton and other macro-invertebrates the chances of scarcity of fish food is higher in Monsoon months. This may be a limiting factor for the growth of fishes in addition to the other anthropogenic activities, which disturb and destruct the fish ecology. Sivakumar (2003) made observations on some biological aspects of the fresh water fish. During the present study period different species of fish fauna were collected from the different sites of Henwal River. Fishes were collected from various spots of River Henwal, all through the sampling range. The river contained rich fish diversity and population.

The problem of the degrading water of the river has not just come up but is on the way to take serious dimensions in certain stretch. Now is the high time to think and take steps to conserve this rich resource before it faces a major alteration which may not only influence the human but may induce either more adaptability of the biota living in it or may cause partial damage to the various species which would not be able to adapt to the fast change. Some may even face forced shifting from their natural habitat, endangering them or may even cause death of the sensitive flora and fauna.

The river Henwal is used by the villagers for several purposes. The study on the river is scanty till now, so that it is important to study the water of river Henwal. The present study will be done at a stretch of about 15 Kms. The samling station B i.e. Khari is situated about 7 Km far from sampling station A and Sampling station C is situated at a distance of about 8 Kms. from samplig station B. It is necessary to monitor the water quality of river Henwal regularly to know the limnological characterization and its relation to biodiversity.

REFERENCES

Abdin, G. (1948a): Physical and chemical investigations relating to algal growth in river Nile. *Cairo. Bull. Inst. Egypt*, 29:20-24.

Abdin, G. (1948b): Seasonal distribution of Phytoplankton and sessile algae in the river Nile. *Cairo. Bull. Inst. Egypt*, 29:369-382.

Abohweyere, P.O. (1990): Study of limnological parameters and potential fish yield in Kigera reservoir (Extensive system) in Kainji, New Bussa, Nigerie. *J. Aqua. Sci.,* 5: 53-58.

Adoni, A.D. (1985): Eutrophication and production studies of some lentic ecosystem around Sagar (M.P.) Ann. Rept. MBA Proj., 100-108.

Ahmed, N. (1952): Stocking of fish in East Pakistan. *Agric. Pakist.*, 3 (4):1-3.

Ahmed, K and Begum, J. (2002): Bacteriological quality of water in hill stream of Khanpara during different seasons. *Nature Environment and Pollution Technology,* 1(2), 171-173.

Akopian, M., Garnier, J. and Pourriot, R. (1999): A Large reservoir as a source of zooplankton for the river structure of the populations and influence on fish predation. *J. Plankt. Res.,* 21(2): 285-297.

Allen, J.R. L. (1985): Principles of physical sedimentology. London. Allen & Unwin.

Armantrout, N.B. (1999): Glossary of aquatic habitat inventory technology. American Fisheries Society., 150 pp.

Arunachalam, M. (2000): Assemblage structure of stream fishes in the western Ghats (India). *Hydrobiologia.,* 430: 1-31.

Arunachalam, M., Johnson, J. A. and Sankaranarayanan, A. (1997a): Fish diversity in rivers of Northern Karnataka. *Int. J. Ecol. Envir. Sci.,* 23: 327-333.

Arunachalam, M., Johnson, J.A., Soranam, R. and Haniffa, M.A. (1997b): Fish diversity in Chittar river of Western Ghats. *Int. J. Ecol. Envir. Sci.,* 23: 335-342.

Arunachalam, M., Nair, K.C., Vijverberg, J. and Kartmulder, K. (1988): Food and habitat usage of a cyprinid fish assemblage in stream pools of a south indain River. Rep. 19. Limnological Institute of the Royal Academy of Science. The Netherlands., 89pp.

Bachmann, R.W. (2001): The Limiting Factor Concept: What Stops Growth? Lakeline, Spring 2001: 26-28.

Badola, S. P. and Singh, H.R. (1981): Fish and fisheries of river Alaknanda *Proc. Nat. Acad. Sci.,* 15(B):133-142.

Badoni, K., Nautiyal, R., Bhatt, J.P., Kishor, B. and Nautiyal, P. (1997): Variations in the epilithic diatom community structure due to river valley projects on the Ganga between Rishikesh and Haridwar. *Proc. Indian Natn. Sci. Acad.,* B 68 No. 6. pp 523-536.

Badola, S.P. and Badola, Smita (1999): Observations on destructive fishing methods and recommendations to protect fish fauna in Garhwal Himalaya, *Sus. Eco. SysEnv. &*, Ed. By D.R. Khanna, A.Gautam and A.Gautam pp-99-104.

Bahuguna, S. N. and Badoni, A. K. (2002): Qualitative and quantitative productivity of fish food from three important tributaries of river Alaknanda, Garhwal. *Him. J. Env. Zool.,* Vol.16(2), pp.215-222.

Bailey, R.G. (1995): Description of the ecoregions of the United States (2nd edn.). Miscellaneous publication no. 1391. Map scale 1:7, 500,000. U.S. Dept. of Agriculture, Forest Service.

Bain, M.B., Finn, J.T. and Booke, H.E. (1985): Quantifying stream substrate for habitat analysis studies. *N. Am. J. Fish. Manage.,* 5: 499-500.

Baker, J.A., Killgore, K.J. and Kasul, R.L. (1991): Aquatic habitats and fish communities in the lower Mississipi River. *Reviews in Aquatic Sciences.,* 3: 313-356.

Barbara, Bis Lambertus and Higher, W.G. (2001): Riparian vegetation of stream and the macroinvertebrate community structure. *Ecohydrol. Hydrobiol.,* 1: 253-260.

Barbour, M.T., Gerritsen, J. and White, J.S. (1996): Development of the Stream Condition Index (SCI) for Florida. Tetra Tech. Inc., Owings Mills, MD.

Bass, D. (1994): Community structure and distribution patterns of aquatic macroinvertebrates in a Tall Grass Prairie stream ecosystem. Proc. Okla. Acad. Sci., 74: 3-10.

Basu, M. and Mitra, A.K. (2002): Studies on the pollution of river Damodar at Burdwan with special reference to metals, COD and pH, *J. Nat Env.& Poll Tech.,* 1(4): 397-399.

Belsare, D.K., Gautam, A., Prasad, D.Y. and Gupta, S.N., (1990): Limnological studies on Bhopal lakes. Seasonal changes in abiotic factors in the upper lake. *Proc. Net. Acad. Sci.,* 60: 147-150.

Benson, A.C., Sutton, T.M., Elliott, R.F. and Meronek, T.G. (2005): Seasonal movements patterns and habitat preferences of age-0 lake sturgeon in the lower Peshtigo River, Wisconsin. *Trans. Am. Fish. Soc.,* 134: 1400-1409.

Berner, L.M. (1951): Limnology of lower Missouri river, *Ecology,* 32: 1-12.

Beschta, R.L. and Platts, W.S. (1987): Morphological features of small streams: significance and function. *Water Resources Bull.,* 22: 369-379.

Bhat, A. (2003): Diversity and composition of freshwater fishes in river systems of central Western Ghats, India. *Environ. Biol. Fish.,* 68: 25-38.

Biggs, B.J., Smith, R.A. and Duncan, M.J. (1999): Velocity and sediment disturbance of priphyton in headwater streams: Biomass and metabolism. *J. N. Am. Benthol. Soc.,* 18(2): 222-241.

Bilby, R.E. and Ward, J.W. (1989): Changes in Characteristics and function of woody debris with increasing size of streams in Western Washington. *Trans. Am. Fish. Soc.,* 118: 368-378.

Bilgrami, K.S. and Dutta Munshi, J.S. (1985): Ecology of river Ganges (Patna-Farraka); Final Technical Report. Pp. 91 (Bhagalpur: Allied Press).

Bisson, P.A. and Montgomery, D.R. (1996): Valley segments stream reaches and channel units. Pp. 23-42. In: R.F. Haur and G.A. Lambert (Eds.). Methods in stream ecology. *Academic Press, San Diego*, California.

Bisson, P.A., Sllivan, K. and Nielsen, J.L. (1988): Channel hydraulics, habitat use and body form of juvenile Coho salmon, steelhead and cutthroat trout in steams. *Trans. Am. Fish. Soc.,* 117: 262-273.

Bilgrami, K.S. and Duttamunshi, J.S. (1985): Ecology of river Ganges (Patna-Farakka). Technical Report, CSIR.

Bisht, R.S. and Das, S.M. (1985): Abiotic parameters and aquatic insects of two cold water Kumaon lakes. *Indian J. Ecol,* 12 (2) : 365-369.

Bomann, F.H., Lekens, G.E. and Easton, J.S. (1969): Biotic regulation of particulate and solution losses from a forest ecosystem. *Bioscience.,* 19: 600-611.

Bovee, K.D. and Milhous, R.T. (1978): Hyraulic simulation in instream flow studies: Theory and techniques. Instream flow information paper No. 5. Washington DC: US Fish and Wildlife Service. OBC78 (33):1-131.

Bowen, Z.H., Freeman, M.C. and Bovee, K.D. (1998): Evaluation of generalized habitat criteria for assessing impacts of altered flow regimes on warm water fishes. *Trans. Am. Fish. Soc.,* 127: 455-468.

Boyero, L. and Bosch, J. (2004): The effect of riffle-scale environmental variability on macro-invertebrate assemblage in a tropical stream. *Hydrobiologica.*524:125-132.

Braaten, P.J. and Berry, C.R. Jr. (1997): Fish association with four habitat types in a South Dakota Praire stream. *J. Freshwat. Ecol.,* 12(3): 477-489.

Bretschko, G. and Waidbacher, H. (2001): Riparian ecotones, invertebrates and fish: life cycle timing and trophic base. *Ecohydrol. Hydrobiol.,* 1: 57-64.

Bridge, J.S. (2003): Rivers and Floodplains. *Oxford Blackwell Sci.* 491pp.

Brune, G.M. (1953): Trap Efficiency of Reservoirs. Transactions of the American Geophysical Union, 34 (3): 407-418.

Bustard, D.R. and Narver, D.W. (1975): Aspects of the winter ecology of juvenile Coho salmon *(Oncorhynchus kisutch)* and steelhead trout *(Salmo gairdneri). J. Fish. Resea. Board of Canada.,* 32: 667-680.

Cairns, J., Jr. (1974): Protozoan (Protozoa). 1-28 pp. In: C.W. Hart and L.H. Fuller (eds.) Pollution Ecology of Freshwater Invertebrates. *Academic Press.* New York.

Cairns, J., McCormick, P.V. and Niederlehner, B.R. (1993): A Proposed framework for developing indicators of ecosystem health. *Hydrobiologia.,* 263: 1-44.

Canfield, D.E., and Bachmann, R.W. (1981): Prediction of Total Phosphorus Concentrations, Chlorophyll *a*, and Secchi Depths in Natural and Artificial Lakes. *Canadian Journal of Fisheries and Aquatic Sciences* 38: 414-423.

Carlson, R.E. (1977): A Trophic State Index for Lakes. Limnology and Oceanography 22: 361-369.

Chacko, P.I. and Ganpati, S.V. (1949): Some observations on the Adyar river with special reference to its hydrobiological conditions *Indian Geogr. J.,* 24 (3)

Chacko, P.I. and Ganpati, S.V. (1952): Hydrobiological survey of the Surli river in the highway of the Madurai district, India, to determine its suitability for the introduction of rainbow trout. *Arch. Hydrobiol.*, 48:128-141.

Chacko, P.I. and Srinivasan, R. (1955): Observations on the hydrobiology of the major rivers of Madras state, south India, Fresh Biol. Stn. Madras, 13:1-14.

Chakraborty, R.D., Roy, P. and Singh, S.B. (1959): A quantitative study of plankton and physico-chemcial conditions of river Yamuna at Allahabad in 1954-55. *Indian J. Fish.,* 61: 186-208.

Chandler, D.S. (1940): Limnological studies of western lake Eric 1: plankton and certain physico-chemical data Bais Island region from September 1938 to November 1939. *Ohio. J. Sci.,* 40: 291-336.

Chandler, D.C. (1944): Limnological studies of western lake Eris. N. Relation of Limnological and climate factors to phytoplankton of 1941. Trans. Amer. Micros. Soc., 63: 203-235.

Chopra, A.K. and Patrik J.N., Nirmal J. (1994): Effect of domestic sewage on self-putrification of Ganga water at Rishikesh. Physico-chemical parameters. *Ad.Bids.* Vol. 13(1): 75-82.

Chugunova, N.I. (1963): Age and growth in fish. National Science Foundation, Washington pp. 132.

Chugh, Tarun (2000): Seasonal variation in the microbial ecology of River Ganga at Haridwar, Ph. D. thesis submitted to Gurukul Kangri University Haridwar.

Close, M.E. and Davies-Colley, R.J. (1990): Baseflow chemistry in New Zealand rivers 2. Influence of environmental factors. *NZ. J. Mar. and Freshwater Res.,* 24: 343-356.

Connell, J.H. (1978): Diversity in tropical rainforest and coral reefs. *Science.,* 199: 1302-1310.

Cooke, G.D. and R.H. Kennedy. (2001): Managing Drinking Water Supplies. Journal of Lake and Reservoir Management 17 (3): 157-174.

Copp, G.H. (1991): Typology of aquatic habitats in the Great Ouse, a small regulated lowland river. *Regulated Rivers: Research and Management.,* 6: 125-134.

Corkum, L.D. (1992): Spatial distributional patterns of macroinvertebrates along rivers within and among biomass. *Hydrobiologia.,* 239: 101-114.

Cuffney, T.F., Gurtz, M.E. and Meadow, M.R. (1993): Methods for collecting benthic invertebrate samples as part of the national water quality assessment program. U.S. Geological Survey Open File Report, Raleigh, NC. Pp 93-406.

Cummins, K.W. (1962): An evaluation of some techniques for the collection and analysis of benthic samples with special emphasis on lotic waters. *Am. Midl. Nat.,* 67: 477-504.

Cummins, K.W. (1992): Invertebrates. In Calow, P. and G.E. Petts (eds.). The River Handbook. Blackwell Science, Oxford: 234-250.

Das, S.M. (1959): Studies on freshwater plankton, III: Quantitative composition and seasonal fluctuations in plankton components. Proc. Nat. Acad. Sci. India., 29: 174-189.

Das, S.M. and Srivastava, V.K. (1956): Quantitative studies in freshwater plankton. II: Correlation between plankton and hydrological factors. Proc. Nat. Acad. Sci. India., 26: 243-254.

Das, S.M. and Moitra, S.K. (1955): Studies on the food of some common fishes of Uttar Pradesh, India. The surface feeders, the mid feeders and the bottom feeders *Proc. Nat. Acad. Sci. India,* 25 B (1 and 2): 1-6.

Datta Munshi, J.S. and Srivastava, M.P. (1988): Natural history of fishes and systematic of freshwater fishes in India. Narendra Publishing Co. Delhi, India. 421 pp.

David, A., Rao, N. G. S. and Rahman, F.A., (1969): Limnology and fisheries of Tungabhadra reservoir. *Bull C.I.F.R.I.* Barrachpore, 13:188.

Davis, W.M. (1899): the geographical cycle. *Geographical journal.,* 14: 481-504.

Day, F. (1875-78): Fishes of India; being a natural history of the fishes known to in habit the seas and freshwaters of India, Burma and Ceylon. Text and Atlas in 4 parts. London: XX + 778, 1956 pls.

Day, J. W. Jr., Hall, C.A.S., Kemp, W.M. and Yanez-Arancibia, A. (1989): Estuarine ecology. John Wiley and Sons, New York. 558 pp.

Depiereux, E., Feytmans, E. and Micha, J.C. (1984): Utilization critique de analyse en composants principales et du cluster analysis pour la description d'ecjamto;;pms d'invetebres benthiques en eau douce. *Oikos.,* 40: 81-94.

De Shon, J.E. (1995): Development and application of the invertebrate community index (ICI). In: Dacus, W.S. & T.P. Simon, (eds.). Biological Assessment and Criteria. Lewis Publishers, England: 217-243.

Dobriyal, A.K. and Singh, H. R. (1981): Diurnal variation in some aspects of limnology of the river Mandakani from the Garhwal Himalaya. *Uttar Pradesh J. Zoo.,* 1:16-18.

Dobriyal, A. K., Singh, H. R. and Bist, K. L. (1983): Diurnal variation in hydrobiological parameters of two hill streams of Garhwal Himalaya . *Uttar Pradesh J. Zoo.,* 3:30-34.

Dobriyal, A.K. (1985): Ecology of limnofauna in the small streams and their importance to the village life in Garhwal Himalaya. *Uttar Pradesh J. Zoo.,* 5(2): 139-144.

Dobriyal, A.K. and Singh, H.R. (1987): A case study on the origin of rhithroplankton in Garhwal Hill streams. *Agri-Biol.* 3(2): 104-106.

Donald, J.O., Jones, R.N. and Maughan, O.E. (1982): Species composition and relative abundance of benthic macroinvertebrates in Glover Creek, Southeast Oklahoma. Proc. Okla. Acad. Sci., 62: 18-21.

Dua, A.and Johal, M.S. (2000): Use of fish scales as an indicator of pesticide pollution. *Proceeding National Symposium on 'Aquaculture for 2000 A.D.'* Organized by Madurai University, Madurai, India. PP. 283-288.

Duffield, R.M. and Nelson, C.H. (1993): Seasonal changes in the stone fly (Plecoptera) components of the diet profile of trout in Big Hunting Creck. Maryland, USA. *Aquat. Insects.,* 15: 141-148.

Dutta, S.P.S., Malhotra, Y.R., Sharma, K.K., and Sinha, K. (1997): Diel variations in physico-chemical characteristics of water and macrobenthic invertebrates in river Tawi, near Nagrota by pass, Jammu. *Him. J. Env. Zool.*, Vol.11, pp.45-50.

Dutta, A., Ahmad, T. Md., Mukherjee, M., Dutta, S. and Pandey, P. (2002): presented an overview of the fish diversity of Ranchi district *Him. J. Env. Zool.* , Vol.16(1), pp.257-262.

Edds, D.R. (1993): Fish assemblage structure and environmental correlated in Nepal's Gandaki River. *Copeia.* PP 48-60.

Eddy, S. (1934): A Study of the fresh water Plankton Communities Illinois, Biol. Monogr, 12 (4): 1-93.

Edmondson, W.T. (1961): Changes in lake Washington following an increase in nutrient income; Verth. Int. Ver. Limnol., 14: 167-175.

Edmondson, W.T. (1977): Population dynamics and secondary production: Ergeb. Limnoi., 56-64.

Edwards, E.D. and Huryn, A.D. (1996): Effect if riparian land use on contribution of terrestrial invertebrates to streams. *Hydrobiological.,* 33: 151-159.

Eklove, P. and Diehl, S. (1994): Piscivore efficiency and refuging prey: the importance of predator search mode. *Oecologia.,* 98: 344-353.

Egborge ABM (1978c): Observations on the diurnal changes in some physico-chemical variables of Lake Asejire, a new impoundment in Nigeria. Pol. Arch. Hydrobiol., 26(2):61-73.

Egborge ABM (1979): The effect of impoundment on the water chemistry of Lake Asejire, Nigeria. Freshwater Biol. 9:403-412.

Egborge ABM (1986): Physical, chemical and biological properties of Benin River. Technical Report for Cage Fishery Development at Kolo Submitted to Nr. J. Ayomike, p. 32.

Esmaeili, H.R, Johal, M.S., Tandon, K.K. and Snarma, M.L. (2001): SEM studied on body scale of a tooth carp, *Aphanius ginaonis* (Holly) from the hot spring of Geno, South Iran. *Abstract Book ,*

XXXIV Annual Conference of EMSI on Electron Microscopy and Allied Fields.pp.42-43.

Evans, J.W. and Noble, R.L. (1979): the longitudinal distribution of fishes in an east Texas stream. *Am. Midl. Nat.,* 101: 333- 343.

Fafioye, O.O., Olurin, K.B. and Sowunmi, A.A. (2005): Studies in the physico-chemical parametes of Omi water body of Ago-Iwoye, Nigeria. *African J. Biotechnology.,* 4(9): 1022-1024.

Fausch, K.D., Lyons, J., Kar, J.R. and Angermeier, P.L., (1990): Fish communities as indicators of environmental degradation. *Am. Fish. Soc. Symp.,* 8: 122-144.

Finger, T.R. (1982): Fish community-habitat reactions in a central. New York stream. *J. Freshwat. Ecol.,* 1: 343-352.

Fjellheim, A. and Raddum, G.G. (1996): Weir building in a regulated west Norwegian river: long term dynamics of invertebrates and fish. Regulated rivers: *Research and Management.,* 12: 501-508.

Flosi, G. and Reynolds, F.L. (1994): California salmonid stream habitat restoration manual. California Dept. of Fish and Game, Technical Report.

Foissner, W. (1988): Taxonomic and nomenclature of Sladeceakis list of ciliates (Protozoa: Ciliophora) as indicator of water quality. *Hydrobiologia.* ,166: 1-69.

Fokmare, A.K. and Musaddique, Mohmmad (2002): A study on the physico-chemical characteristics of kapashi lake and Purna river waters in Alok district of Maharashtra, India. *Nature Environment and Pollution Technology,* 1(3), pp. 261-263.

Forel, F.A. (1892): Le Leman. *Monographic Limnogique*, 3 Vol. Lavsane.

Fraser, D.F. and Cerri. R.D. (1982): Experimental evaluation of predation-prey relationships in a patchy environment: consequences for habitat use patterns in minnows. *Ecology.,* 63: 307-313.

Fuma, S., Ishii, N. and Takedo, H. (2003): Ecological effects of various toxic agents on the aquatic microcosm in comparison with acute ionizing radiation. *J. environ. Radiat.,* 67: 1-14.

Ganpati, S.V. (1940): The ecology of a temple tank containing a permanent bloom of *Microcystis aexiquinosa. J. Bombay N. Hist. Soc.,* 42: 65-67.

Ganpati, S.V. and Chacko, P.I. (1951): Investigation on plankton production and limnology of four fish ponds in Madras. Proc. Int. Assoc. Theor. Applied Limnol., 11:111-112.

Gaufin, A.R. and Tarzwell, C.M. (1952): Aquatic invertebrates from hyporheic sediments of the North Canadian River. Proc. Okla. Acad. Sci., 72: 3-4.

Gautam A. (1988): Limnological studies and seasonal variations in the planktonic organisms and primary production in the upper lake of Bhopal. Ph.D. thesis Bhopal University, Bhopal 140 pp.

Gautam, A., Sarkar, P. and Khanna, D.R. (2001): Quality deterioration in aquatic environment of the Ganges: A growing problem for future. *Indian J. Env. Sci.,* 5(2) pp.113-117.

Gautam, A., Khanna, D.R. and Sarkar, P. (2000): Diurnal variation in the physico-chemical characteristics of Ganga water at Triveni ghat, Rishikesh during winter season. *Indian J. Environ. & Ecoplan.,* 3 (2): 369-371.

George, M.J. (1958): Observations on the plankton of the Cochin backwaters. *Ind. J. Fish.,* 5(2): 375-401.

George, J.P., (1976): Hydrobilogical studies on the lower lake of Bhopal with special reference to the productivity of economic fishes. Ph.D. thesis Bhopal University, Bhopal.

Gorman, O.T. (1988 a): An experimental study of habitat uses in an assemblage of Ozark-minnous. *Ecology.,* 69: 1239-1250.

Gorman, O.T. (1988 b): the dynamics of habitat use in a guild of Ozark minnows. *Ecol. Monogr.,* 58: 1-18.

Gorman, O.T. and Karr, J.R. (1978): Habitat structure and stream fish communities. *Ecology. ,* 59: 507-515.

Gower, A.M., Myres, G., Kent, M. and Foulkes, E. (1994): Relationships between macroinvertebrates communities and environmental variables in metal contaminated streams in south-west England. *Freshwat. Biol.,* 32: 199-221.

Gregary, S.V., Swann, F.J., Mckee, W.A. and Cummins, K.W. (1991): An ecosystem perspective of riparian zones. *Biosciences.,* 41: 450-551.

Griffith, R.S. (1955): Analysis of plankton yields in relation to certain physical and chemical factors of lake Michigan. *Ecology*, 36 (4): 543-552.

Griffith, M.B. and Perry, S.A. (1991): Leaf pack processing in two Appalachian mountains streams draining catchments with different management histories. *Hydrobiologia.,* 230: 247-254.

Goldman, C.R. and Wetzel, R.G. (1963): A study of primary productivity of clear Lake, Lake County, California. Ecology, 44: 283-294.

Grossman, G.D., Moyle, P.B. and Whitaker Jr., J.O. (1982): Stochascity in structure and functional characteristics of Indian stream fish assemblage: a test of community theory. *Am. Natur.,* 120: 423-454.

Hamilton, B. (1822): An account of the fishes found in the river Ganges and its branches. Edingargh and London, VII + 405 pp.

Hamilton, K. and Bergerson, E.P. (1984): Methods in estimate aquatic habitat variables. Colorado Cooperative Fishery Research Unit. 201 Wager. Colorado State University, Fort Collins, CO 80523. Page Var.

Haque, N., Khan, A.A., Fatima, M. nad BArbhuyan, S.I. (1988): Impact of some ecological parameters on rotifer population in tropical perennial pond. *Environ. And Ecol.,* 6(4): 998-1001.

Harrel, R.C. and Dorris, T.C. (1968): Stream order, morphometry, physico-chemical conditions and community structure of benthic macroinvertebrates in an intermittent stream system. *Am. Midl. Nat.,* 80: 220-251.

Harris, J.H. (1995): The use of fish in ecological assessments. *Australian Journal of Ecology.,* 20: 65-80.

Hawkinds, C.P., Hogne, J.N., Decker, L.M. and Feminella, J.W. (1997): Channel morphology, water temperature and assemblage structure of stream insects. *J.N. Am. Benthol. Soc.,* 16(4): 728-749.

Hawkinds, C.P., Kershner, J.L., Bisson, P.A., Bryant, M.D., Decker, L.M., Gregory, S.V., McCullough, D.A., Ovrton, C.K., Reeves, G.H., Steedman, R.J. and Young, M.K. (1993): A hierarchical approach to classifying stream habitat feature. *Fisheries.,* 18: 3-12.

Heede, B.H. and Rinne, J.N. (1990): Hydrodynamic and fluvial morphologica processes: Implications for fisheries management and research. *N. Am. J. Fish. Manage.,* 10: 249-268.

Hoffman, L. (1996): Geographic distribution of freshwater blue-green algae. *Hydrobiologia.,* 336: 33-40.

Holden, J.M. and Green, J. (1960): Hydrobiology and plankton of river Sokkoto. *J. Anim. Ecol.,* 29 (1):65-84.

Hora, S.L. (1921): On some new or rare species of fishes from Eastern Himalayas. *Rec. India Mus.*, 22/5: 731-744.

Hora, S.L. (1940): The game fishes of India VIII. The Mahseers or the large scaled barbus of India. 2. the *Tor* mahseer *Barbus (Tor)* Tor (Hamilton). *J. Bombay Nat. Hist. Soc.,* 41: 518-525.

Hora, W.S. (1957): *Physiology of fishes*. Acad.Press Inc., New York. pp.640.

Hormer, R.R., Welch, E.B., Selley. M.R. and Jacoby, J.M. (1990): Responses of periphyton to changes in current velocity, suspended sediment and phosphorus concentration. *Freshwat. Biol.,* 24: 215-232.

Horowitz, R.J. (1978): Temporal variability patterns and the distributional patterns of stream fish. *Ecol. Monogr.,* 48:307-321.

Huet, M. (1949): Apercu des relations entre la pente et les populations piscicoles des eaux courantes. *Schweiz Z, Hydrol.,* 11: 332-351.

Huet, M. (1972): Textbook of fish culture. Breedingand cultivation of fish . Translated by H. Kohn, Fishing New Book Ltd. Faraham, Survey England. 436.

Hughes, F.M.R. (1997): Floodplain biogeomorphology. *Progress in Physical Geography.,* 21: 501-529.

Hussaini, S.U. (1967): Studies on the limnological and primary production of tropical lake. Hydrobiologia. 30 (3-4): 335-352.

Hutchinson, G.E. (1967): A treatise on limnology. Vol II. Introduction to Lake biology and the Limnoplankton. John wiley & Sons Inc. New York, 1115 pp.

Hynes, H.B.N. (1960): the Biology of polluted waters. Liverpool University Press, Liverpool.

Hynes, H.B.N. (1970): In: The ecology of running waters, Liverpool University Press, Liverpool. 543.

Iversen, T.M., Thorup, J., Kjeldsen, K. and Thyssen, N. (1991): Spring bloom development of microbenthee algae and associated invertebrates in two reaches of a small lowland stream with contrasting sediment stability. *Freshwat. Biol.,* 26: 189-198.

Jack, R.G., Maas, J.L. and Scholten, M.C. (1996): Evaluation of laboratory derived toxic effects concentrations of a mixture of metals by testing fresh water plankton communities in enclosures. *Wat. Res.,* 30: 1215-1227.

Jayaram, K.C. (1974): Ecology and distribution of Freshwater fishes, amphibian and reptiles. PP. 517-584. In: M.S. Mani (eds.) Ecology and Biogeography in India. Dr. W. Junk. B.V. Publ. the Hague.

Jena, B., Sudarshana, R. and Chaudhuary, S.B. (2003): Studies on water quality parameters around Sagar island, Sundarbands . *Nature Environment and Pollution Technology,* Vol.2 No.3 pp. 329-332.

Jenkin, P.M. (1942): Seasonal Changes in the temperature of Windermere (English Lake District). J. Anim. Ecol., 11: 248-269.

Jindal, R. and Singh, H. (2005): Ecology of benthos of some fresh water systems of Punjab. *Proceedings of National Seminar 'New trends in fish development in India',* Punjab University. pp. 179-184.

Johal, M.S., Tondon, K.K., Rawal, Y.K., Tyor, A.K., Banyal, H.S. and Rumana, H.S. (2001): Species richness of fish in relation to environmental factors. *Current Science.,* 80 (4): 499-501.

Johal, M.S., Tondon, K.K., Tyor, A.K. and Rawal, Y.K. (2002): Fish diversity in different habitats in the streams of lower Middle Western Himalayas. *Polish Journal of Ecology.,* 50(1): 45-56.

Johal, M.S., Armantrout, Neil. B. and Arunachalam, M.(2005): Changing Indian Rivers. *Proceedings of National Seminar 'New trends in fish development in India'*, Punjab University. pp. 35-41.

Johal, M.S. and Rawal, Y.K. (2005): Key to management of the Western Himalayan hillstreams in relation to fish species richness and diversity. *Hydrobiologia.*, 532: 225-232.

John, Varkey. (1971): Hydrobiological studies on the river Kallayi in Kerela. Ph.D. thesis of the University of Kerela.

Johnson, L.B., Richards, C., Host, G.E. and Arthur, J.W. (1997): Landscape influences on water chemistry in mid-western stream ecosystems. *Freshwat. Biol.*, 37: 193-208.

Jones, G.P. (1991): Post recruitment process in the ecology of coral reef fish populations: a multi-factorial. PP. 294-328. In: P.F. Sale (ed.). The ecology of fishes in coral reefs. Academic Press, New York.

Kala, Rakhee and Sharma, R.C. (2002): Effect of Physico-chemical factors on phytoplankton population in lotic environment of Alaknanda river, Garhwal Himalaya, Indian *J. Ecol.*, 29(2): 221-226.

Karr, J.R. (1991): Biological integrity: a long neglected aspect of water resource management. *Ecol. Appl.*, 1: 66-84.

Karr, J.R. and Dudley, D.R. (1981): Ecological perspective on water quality goals. *Envir. Manage.*, 5: 56-68.

Katano, O., Anonuma, Y., Nakamura, T. and Yamamoto, S. (2003): Indirect contramensalism through trophic cascades between two omnivorous fishes. *Ecology.*, 84: 1311-1323.

Katano, O., Nakamuram T. and Yamamo, S. (2006): Intraguild indirect effects through trophic cascades between stream dwelling fishes. *Journal of Animal Ecology.*, 75: 167-175.

Kaur, Rajendra and Khulbe, R.D. (1998): Occurrence, Distribution and ecology of watermolds in Nanaksagar reservoir Kumaun Himalaya Nainital, Uttarpradesh India, J. *Indian bot. Soc.*, 17: 111-116

Kaur, H., Bath, K.S., Mander, G. and Dhillon, S.S. (1999): Aquatic invertebrate diversity of Kalanji lake, Punjab. *Indian. J. Env. and Ecoplan.,* 2 (1) : 37-41

Keim, R.F. and Skaugset, A.E. (2002): Physical aquatic habitat II: Pools and cover affected by large woody debris in three Western Pregon streams. *N. Am. J. Fish Manage.,* 22: 151-164.

Kemdirin, E.C. and Ejike, C. (1992): The Phytoplankton response to artificial enrichment with nitrates and phosphates in an upland and lowland reservoir in Plateau state, Nigeria. *J. Aqua. Sci., &:* 45-57.

Keup, L.E., Ingram, W.M. and MacKenthum, K.M. (1986): The role of bottom-dwelling macrofauna in water pollution investigations. US Dept. of Health, Educ. And Welfare, Cincinnati, Ohio. 23pp.

Khan, M.A. and Zutshi, D.P. (1980): Primary Productivity and trophic status of a Kashmir himalaya lake. Hydrobiologia., 68 (1): 3-8.

Khanna, D.R. (1993): Ecology and pollution of Ganga River, *Ashish Publication House, Delhi*, 1-241.

Khanna, D.R., Badola, S.P. and Dobriyal, A.K. (1993): Plankton ecology of the river Ganga at Chandighat, Haridwar, *Advances in limnology,* Ed. By H.R. Singh, *Narendra Publishing house,* New Delhi. 171-174.

Khanna, D.R. and Bhutiani, R. (2007): Limnological Modeling: A case Study of River Suswa. *Daya Publishing House*: 1-297.

Khanna, D.R. and Bhutiani, R. (2005): Water analysis- At a Glance. ASEA publication, 1-116.

Khanna, D.R. and Bhutiani, R. (2004): Fish and their ecology of River Ganga at Gohri Ghat Garhwal (Uttaranchal), *Proc. Nat. Sem,* Fish diversity in protected habitats, Ed. By S. Ayyappan, D.S. Malik, Rani Dhanze and R.S. Chauhan: 291-298.

Khanna, D.R. and Bhutiani, R. (2003): Limnological characteristic of the river Ganga at Haridwar (Uttaranchal), *U. P. J. Zool.,* 23 (3): 179-183.

Khanna, D.R. and Bhutiani, R. (2003): Limnological status of Satikund pond at Hardwar (U.A.), *Indian J. Env. Sc.*, 7 (2): 131-136.

Khanna, D. R., Bhutiani, R., Chandra, K.S. (2005): Modeling to assess effect of light on phytoplanktonic growth dynamics: A Review. *Environmental Conservation Journal*. 6 (1-3): 103-107.

Khanna, D. R. and Chugh, T. (2004): Microbial Ecology: A study of River Ganga, *Discovery publishing house*, New Delhi. 1-277.

Khanna, D.R. Chugh, Tarun and Sarkar, Praveen (2001): Fluctuation in the population density of Macro-invertebrates of river Ganga at Pashulok barrage, Rishikesh (Uttaranchal, India), *Env. Cons. J.*, 2 (1): 37-39.

Khanna, D.R., Gautam, Ashutosh, Chugh, Tarun and Sarkar, Praveen (2000): Impact of abiotic factors on the phytoplanktonic population of a pond, *Env. Con. J.*, 1(1): 41-46.

Khanna, D.R. Malik, D.S. and Rana, D.S. (1997): Phytoplanktonic communities in relation to certain physico-chemical parameter of Ganga canal at Haridwar., *Him. J. Env. Zool.*, 12: 193-197.

Khanna, D.R., Malik, D.S., Seth, T.R. and Rupendra (1999): Correlation Between abiotic factors and planktonic population in river Ganga at Rishikesh (U.P), *Sus. Eco Sys. and Env*, Ed. by D.R.Khanna, A Gautam and A. Gautam, Published by ASEA, Rishikesh: 69-76.

Khanna, D.R., Malik, D.S. and Rupendra (1998): Fish fauna of the river Ganga at Rishikesh. *J. Natcon.* 10 (2): 212-213.

Khanna, D.R., Pathak, S.K., Bhutiani, R. and Chandra, K.S. (2006): Study of water quality of River Suswa Near Raiwala, Uttaranchal. *Env. Con. J.* 7 (3):79-84.

Khanna, D.R., Seth, T.R., Malik, D.S. and Singh, R.K. (1999): Fishes and their ecology of river Suswa at Raiwala (Dehradun), *Sus. Eco. Sys. and Env.*, Ed. By D.R. Khanna, A.Gautam and A.Gautam: 105-110.

Khanna, D.R. and Singh, S. (2002): Pond fish Ecology and Economics, *Daya Publishing House*, Delhi 1-816.

Khanna, D.R. and Singh R.K. (2000): Seasonal fluctuations in the plankton of Suswa river at Raiwala (Dehradun)., *Env. Cons. J.*, 1 (2 and 3): 89-92.

Khanna, D.R. and Singh, Shekhraj. (1999): Diurnal variation in microbial ecology of Narkewala Pond at Ferupur, Hardwar, *Sus. Eco. Sys. and Env.*, Ed. By D.R. Khanna, A.Gautam and A.Gautam: 99-104.

Khanna, D.R., Singh, Vikas, Bhutiani, R., Chandra, Kumar Satish, Matta, Gagan and Kumar, Dheeraj (2007): "A Study of biotic and abiotic factors of Song River At Dehradun, Uttarakhand". *Env. Conservation J.* 8 (3)-117-126.

King, D.L. (1970): The role of carbon in eutrophication. *J. Wat. Pollut. Control Fed.,* 42: 2035-2051.

King, R.P. (1988): Physico-chemical indices of the fisheries potential of a Nigerian rainforest Pond. *J. Aqua. Sci.,* 13: 49-54.

Kobayashi, T., Shiel, R.J., Gibbs, P. and Dixon, P.L. (1998):Freshwater zooplankton in the Hawkesbury Nepean River: comparison of community structure with other rivers. *Hydrobiologia.,* 337: 133-145.

Koetsier, P., Minshall, G.W. and Robinson, C.T. (1996): Benthos and macroinvertebrate drift in six streams differing in alkanity. *Hydrobiologia.,* 317: 41-49.

Koteswari, Y.N. and Ramanibai, R. (2003): Impact of tannery effluent in the colonization of planktons: a microcosm study. *Turk. J. Biol.,* 27: 163-170.

Kreshner, J.L. and Sinder, W.M. (1992): Importance of a habitat-level classification system to design instream flow studies. In: P.J. Boon, P. Calow and G.E. Petts (eds.). River Conservation and Management, New York: John Wiley and sons. PP 179-193.

Kudesia, V.P. and Verma, S.P. (1985): Industrial pollution on Kali river *J. Env. Sci.,* 1(2): 41-49.

Kuehne, R.A. (1962): A Classification of streams, illustrated by fish distribution in an eastern Kentucky creek. *Ecology.* , 43: 608-614.

Kumar, Arvind (1995): Studies in pollution in river Mayurakshi in South Bihar. *Indian J. Env. Poll.* 2(1) 21-26.

Kumar, Neeraj and Sharma, R.C. (2002): Water quality of river Krishni {Part-I Physicochemical characteristics}, *J.Nat.Con.,* 14(2): 273-297.

Kumar, Neeraj and Sharma, R.C. (2002): Water quality of river Krishni {Part-II Physicochemical characteristics}, *J.Nat.Con.,* 14(2): 299-333.

Kumar, A. and Bohra, R. (1999): Gastropods as indicators of pollution status of some wetlands in Santal pargana, Bihar, India. *Indian J. Env. and Ecoplan.,* 2(1): 83-87.

Kumar, S. Sathya and Mukherjee, S.K. (2000): *Recent changes in the flora and fauna of Doon Valley and surrounding area.* A report submitted to the Supreme Court Monitoring Committee. W.I.I., March 2000, pp. 27-45.

Kushlan, J.A. (1976): Environmental stability and fish community diversity. *Ecology.,* 57:821-825.

Laal, A.K., Sarkar, S.K. and Sarkar, A. (1988): Ecology and fisheries of River Ganga at Bhagalpur (Bihar). Proc. Natl. Sym. Fish. Env.: PP. 51-55.

Lamb, M.A. and Lowe, R.L. (1981): Primary productivity and capital structure of phytolithic growth in streams of the Great Smoky Mountains, National Park Tennesee. *Hydrobiologia.,* 123: 59-67.

Lamberti, G.A. (1996): The role of periphyton in benthic food webs. In: R.G. Stevenson *et al.* (eds.), Algal ecology. Academy Press, San Diego, CA: 533-573.

Lawrence, R.R., Leonard, R. and Sheridan, J. (1985): Managing riparian ecosystems to control nonpoint pollution. *J. Soil and water Conserv.,* 40: 87-91.

Leveque, C. (1997): Biodiversity dynamics and conservation. The freshwater fish tropical Africa, Cambridge University Press, Cambridge, 438 pp.

Lewis, W.M. (1974): Primary production in the plankton community of a trophical lake. Ecol. Monogr. 44 (4): 337-409.

Likens, G.E., Bormann, F.H., Johnson, N.M., Fisher, D.W. and Pierce, R.S. (1970): Effects of forests cutting and herbicide treatment on nutrient budgets in the Hubbard Brook catchment ecosystem. *Ecol. Monogr.,* 40: 87-91.

Lindman, R.L. (1942): The Trophic dynamic aspect of ecology. Ecology. 23:399-418.

Lloyd, R. (1961): Study on the toxicity of mixture of zinc and copper sulfates to rainbow trout (*Salmo gaidnerii*). Richardson. *Ann. App.Biol.,* 49: 535-538.

Lotrich, V.A. (1973): Growth, production and community compositon of fishes inhabitating a first, second and third order stream of eastern Kentucky. *Ecol. Monofr.,* 43: 377-397.

Lund, J.W.G. (1972): Eutrophication. Poc. R. Soc. London Ser. B.,180: 371-382.

Mackereth F.J.H. (1963): Some methods of water analysis for limnologists. Freshwater Biol. Ass, Sci. Pub. 21: 72.

Mahajan, C.L. (1961): Fish fauna of Muzzafarnagar district Uttar Pradesh. *J. Bombay Nat. Hist. Soc.,* 62 (3): 440-454.

Matthews, W.J. and Hill, L.G. (1980): Habitat partitioning of fish community of a southwestern river. *Southwest Nat.,* 25: 51-66.

Mazzoni, R. and Iglesias-Rios. R. (2002): Distribution pattern of two fish species in a coastal stream in southeast Brazil. *Braz. J. Biol.,* 62 (1): 171-178.

McCledon, D.D. and Rabeni, C.F. (1987): Physical and biological variables useful for predicting population characteristics of the small mouth bass and rock bass in an Ozark stream. *N. Am. J. Fish. Manage.,* 7:46-56.

McGehee, M.A. (1994): Correspondence between assemblages of coral reef fishes and gradients of water motion, depth and substrate size, off Puerto Rico. *Mar. Ecol. Prog. Scr.,* 105: 243-255.

Merilaita, S. (2003): Visual background complexity facilities the evolution of camouflage. *Evolution.,* 57: 1248-1254.

Meyer, J.L. and Johnson, C. (1983): The influence of elevated nitrate concentration on the rate of leaf decomposition in a stream. *Freshwat. Biol.,* 13: 177-183.

Meyer, J.L., McDowel, W.H., Both, T.L., Elwood, J.R., Ishizaki, C., Melack, J.M., Peckarsky, B.L., Peterson, B.J. and Rublee, P.A. (1988): Elemental dynamics in streams. *J. N. Amer. Benthol. Soc.,* 7: 410-432.

Michael, P. (1984): In: Ecological method for field and laboratory investigation. Tata Mc-Graw Hill Publi. Ltd, New Delhi.

Minshall, G.W. and Minshall, J.N. (1977): Microdistribution of benthic invertebrates in a Rocky mountain (USA) stream. *Hydrobiologia.,* 55: 231-249.

Moitra, S.K. and Bhowmik, M.L. (1968): Seasonal cycles of rotifers in a fresh water fish pond in Kalyan, west Bengal. Proc. Symp. Adv. Trop. Ecol., Part I: 359-367.

Monichan, K.K. and Sharma, R.K. (2003): Distribution status of fish fauna in Etawah district, U.P. *Indian J. Env. Sciences,* 7(2), 115-117.

Monizsette, D.G. and Mavinic, D.S. (1978): BOD Test Variables. J. Envir. Eng. Div., EE6: 1213-1232.

Montgomery, D.R., Buffington, J.M., Smith, R.D., Schmidt, G.M. and Press, G. (1995): Pool Spacing in forest channels. *Water Resources Research.,* 31: 1097-1105.

Moore, K., Jones, K. and Dambacher, J. (1995): Methods of stream habitat surveys, version 5.1 Oregon Dept. of Fish and Wildlife, Corvallis.

Motwani, M.P., Banerji, S.M. and Karam Chandani, S.J. (1956): Some observation on the pollution of river Sone by the factory effluent of Rohtas industries at Dalmia nagar (Bihar). *Indian J. Fish.,* 3:334-367.

Moyle, P.B. and Senanayake, F.R. (1984): Resource partitioning among the fishes of rainforest streams in Sri Lanka. *J. Zool. Soc. Lond.,* 202: 195-224.

Nautiyal, P. and Lal, M.S. (1982 a): Fecundity of Garhwal Himalayan Mahseer, *Tor putitora* (Ham.) *J. Bombay Nat. Hist. Soc.,* 82: 253-257.

Nautiyal, P. and Lal, M.S. (1982 b): Recent record of Garhwal Himalayan Mahseer (*Tor putitora*) with a note on its present status. *J. Bombay Nat. Hist. Soc.,* 79: 639-695.

Nautiyal, P. (1984): Studies on the rivrine ecology of torrential waters in the Indian uplands of the Garhwal region II: Seasonal fluctuations in diatom density. Proc. Indian Acad. Sci., 93 (7): 671-674.

Nautiyal, P., Dobriyal, A.K., Chaukiyal, D.C., and Singh, H.R. (1986): Quality and potability of Alaknanda river at Srinagar- Garhwal (U.P., India) A note on Bacterial density as an indicator of pollution. *Poll. Res.,* 5(3-4): 143-145.

Nautiyal, P. (1986): Studies on the riverine ecology of the torrential waters in Indian uplands of the Garhwal rgion. 3. Flouristic and Faunistic Survey. *Trop. Ecol.,* 27: 125-165.

Nautiyal, P. (1990): Ecology of the Ganga river system in the uplands of Garhwal. In "Himalayas: Env. Resources and Development". N.K. Shah, S.D. Bhatt and R.K. Panda (eds.) Shree Almora Book Depot. Almora: 67-72.

Nautiyal, P. (1996): The river valley projects in Garhwal region. Impact on the population dynamics of endangered Himalayan Mahseer with emphasis on bio-conservation. Final Technical Report of Research Project No. J-1103/8/92-G. PD., submitted to Ganga Project Directorate, Ministry of Environmental and Forests, Government of India, New Delhi, 1996. PP. 1-93.

Nautiyal, P. and Lal, M.B. (1978): Eco-biological studies of some hill stream fishes of Garhwal Himalayas III. Hydro-biological factors and their bearing on the productivity of Mahseer in the hillstreams, Alaknanda, Mandakini and Nayas, *Indian J. Zool.,* 19: 21-27.

Nautiyal, R., Nautiyal, P. and Singh, H.R. (1996): Impact of sewage on the diatom communities of River Alaknanda (Srinagar, Garhwal). *Int. J. Ecol. Envir. Sci.,* 22: 289-296.

Needham, J.G. and Needham, P.R. (1972): *A guide to the study of Fresh Water Biology*. Holden-Day 1. N.C.San Francisco, *Coilt.* 94 (3): 1-108.

Oliff, W.O. (1959): Hydrobiology of Tugela river system. Part I. the main Tugela river *Hydrobiologia*, 14: 281-370.

Omernik, J.M. (1995): Ecoregions: a spatial framework for environmental management. In David, W.S. and T.P. Simon (eds.). Biological Assessment Criteria: Tools for water resource planning and decision making. CRC Press, FL.

Osborne, L.L., and Wiley, M.J. (1988): Empirical relationships between land use/cover pattern and stream water quality in an agricultural catchments. *J. Envir. Manage.,* 26: 9-27.

Paller, M.C. (1994): Relationships between fish assemblage structure and stream order in South Carolina coastal plain streams. *Trans. Am. Fish. Soc.,* 123: 150-161.

Parsons, M. and Norris, R. H. (1996): The effect of habitat-specific sampling on biological assessment of water quality using a predictive model. *Freswat. Biol.,* 36: 419-434.

Pathak, S.K. and Mudgal, L.K. (2005): Limnology and biodiversity of fish fauna in Virla Reservoir, M.P. India, *Env. Con. Jour*. Vol. 6 (1): 41-45 pp.

Pathani, S.S., Upadhya, K.K. and Joshi, S.K. (2002). Some physico-chemical parameters and primary productivity of river west Ramganga. *Him. J. Env. Zool.,* Vol. 16(2), pp. 151-158.

Pearsall, W.H. (1930): A theory on diatom periodicity. Proc. Rev. Soc., 11:105-183.

Pennak, R.W. (1949): Annual Limnological Cycles in some Colorado reservoir and lakes. Ecol. Monogr., 91:233-267.

Phillips, C.P. C. (1927): Diurnal Fluctuation in the hydrogen ion activity of a Minnesota Lake. Ecology, 8:73-89.

Pokhriyal, R.C., Dobriyal, A.K. and Singh, H.R. (1983): Seasonal variation in potamological parameters of the stream Khandagadh. *Uttar Pradesh J. Zoo.*, 3:70-72.

Pires, A.M., Cowx, I.G. and Coelho, M.M. (2000): Benthic macroinvertebrate communities of intermittent streams in the middle reaches of the Guadiana Basin (Portugal). *Hydrobiologia.,* 435: 167-175.

Postma, H. (1967): Sediment transport and sedimentation in the estuarine environment. PP. 158-179. In: G.H. Lauff (ed.). Estuaries, American Association for the Advancement of Science. Publication No. 83, Washington, D.C.

Power, M.E. and Matthews, W.J. (1983): Algae-grazing minnows (*Compostoma anomalum),* piscivorous bass *(Micropterus* spp.*)* and the distribution of attached algae in a small prairie-margin stream. *Oecologia.* 60: 328-332.

Pozo, J., Basaguran, A., Elosegni, A., Molinera, J., Fabre, E. and Chaver, E. (1998): Afforestation with *Eucalyptus globules* and leaf lit decomposition in streams of northern Spain. *Hydrobiologia.,* 373/374: 101-104.

Prasad, R.P. (1956): Further studies on the plankton of the in shore waters of Mandapam. Indian J. Fish., 3:1-42.

Pratt, J.R., Niederleher, B.R., Bowers, N. (1987): Prediction of permissible concentrations of copper from microcosm toxicity tests. *Toxicity assessment: An Internat. Qualt.,* 2: 417-436.

Prati, L., Pavanello, R. and Pesarin, F. (1977): Assessment of surface water quality by a single index of pollution. Water Res., 5: 741-751.

Probst, W.E., Rabeni, K.F., Covington, W.G. and Marteney, R.F. (1984): Resource use by stream dwelling rock bass and the small mouth bass. *Trans. Am. Fish. Soc.,* 113: 283-294.

Pusey, B.J., Arthington, A.H. and Road, M.G. (1983): Spatial and temporal variation in fish assemblage structure in the Mary River, south-eastern Queensland: the influence of habitat of structure. *Environ. Biol. Fish.,* 37:355-380.

Rabeni, C.F. (2000): Evaluating physical habitat integrity in relation to the biological potential of streams. *Hydrobiologia.,* 422/423: 245-256.

Rajan, S. (1963): Ecology of fishes of the rivers Pykara and Moyer (Nilgiris), South India. Proc. Ind. Acad. Sci., 58 (B): 291-323.

Ramanibai, R. and Elayaraja, P. (2001): The effects of toxicant on the colonization rate of plankters- A microcosm study. *Indian J. Environ. Ecoplan.* 5: 417-421.

Rankin, E.T. (1995): Habitat indices in water resource quality assessment. In: Davis, W.S. and T.P. Simon (eds.), Biological Assessment and Criteria. Lewis Publishers, Boca. Raton., FL: 181-214.

Rao, N.G. (1987): Synecology of the lake Rangasagar in relation to the limnology and eutrophication. Ph.D. Thesis Sukhadia Univ. Udaipur.

Rao, K.S.., KArtha, K.N., Gupta, D.R., PAndya, S.S. and Iyer, H.K. (1988): Studies on morphometry and hydrobiology of Gandhi Sagar Reservoir with special reference to its fisheries, *J. Fish Tech.,* 25(1): 21-28.

Ray, P., Singh, S.B. and Sehgal, K.L. (1966): A study of some aspects of the river Ganga and Jamuna at Allahabad (U.P.) in 1958-59. *Proc. Nat. Acad. Sci. India,* 36 B (3): 235-272.

Reid, K.G. (1961): Ecology of Inland waters and Estuaries New York: Reinhold, 1-280.

Resh, V.H. (1995): Freshwater benthic macroinvertebrates and rapid assessment procedures for water quality monitoring in developing and newly industrialized countries. In Davis, W.S. & T.P. Simon (eds.), Biological Assessment and Criteria. Lewis Publishers, England: 167-177.

Reice, S.R. (1985): Experimental distribution and the maintenance of species diversity in a stream community. *Oecologia (Berlin).,* 67: 90-97.

Reice, S.R. (1977): The role of animal association, current velocity and sediment in lotic litter decomposition. *Oikos.,* 29: 357-365.

Rhoads, D.C. and Young, D.K. (1970): The influence of deposit-feeding organisms on sediment stability and community tropic structure. *J. Mar. Res.,* 28: 150-178.

Richards, C. and Host, G. (1994): Examining land use influences on stream habitats and macroinvertebrates: A GIS approach. *Water Resources Bulletin.,* 30(4): 729-738.

Richards, C. and Minshall, G.W. (1992): Spatial and temporal trends in stream macroinvertebrate species assemblages: The influence of watershed disturbance. *Hydrobiologia.,* 241: 173-184.

Richards, C., Johnson, L.B. and Host, G.E. (1996): Landscape scale influences on stream habitats and biota. *CanJ. Fish. Aqua. Sci.,* 53 (Suppl. 1): 295-311.

Rutherford, D.A., Echelle, A.A. and Manghan, O.E. (1992): Drainage-wide effects of timber harvesting on the structure of stream fish assemblages in southeastern Oklahoma. *Trans. Am. Fsh. Soc.,* 121: 716-728.

Ruttner, F. (1963): fundamentals of limnology. English translation by D.G. Frey and F.E.J. Fry. Toronto University Press, Canada. 307 pp (first germen edition in 1940).

Sanderson, H., Boudreau, T.M. and Mabury, S.A. (2003): Impact of perfluorooctonic acid on the structure of zooplankton community in indoor microcosms. *Aqua Toxico.,* 62: 227-234.

Sangu, R.P.S. and Sharma, K.D. (1985): Studies on water pollution on Yamuna river in Agra. *Indian J. Environ. Health.,* 27(3): 257-261.

Schlosser, I.J. (1982 a): Fish community structure and function along two habitats gradients in a headwater stream. *Ecol. Monogr.,* 52: 395-414.

Schlosser, I.J. (1982 b): Trophic structure, reproductive success and growth rate of fishes in a natural and modified headwater stream. *Can. J. Fish. Aqu. Sci.,* 39: 968-978.

Schlosser, I.J. (1987 a): The role of predation in age and size related habitat use by stream fishes. *Ecology.,* 651-659.

Schlosser, I.J. (1987 b): A conceptual framework for fish communities in small warm water streams (In: community and evolutionary ecology of North American Stream fishes. W.J. Mathews, D.C. Heins (Eds.). University of Oklahoma Press, Norman. Pp. 17-24.

Schlosser, I.J. (1991): Environmental variation, life history attributes and community structure of stream fishes implications for environment management assessment. *Envir. Manage.*, 114: 621-628.

Schoener, T.W. (1974): Resource partition in ecological communities. *Science.*, 185: 27-39.

Scott, M.C. and Hall Jr. L.W. (1997): Fish assemblage's indicators of environmental degradation in Maryland coastal plain streams. *Trans. Am. Fish. Soc.*, 126: 340-360.

Sedall, J.R., Reeves, G.H., Hauer, F.R., Sranford, J.A. and Hawkins, C.P. (1990): Roll of refugia in recovery from disturbances; modern fragmental and disconnected river systems. *Envir. Manage.*, 14: 711-724.

Seth, T.R., Khanna, D.R., Gautam, A., Chugh, T. and Sarkar, P. (2000): Temporal trends of phytoplanktonic diversity in the river Ganga at Hardwar. *Him. J. Env. Zool.* Vol. 14, 129-134.

Sehgal, K.L. (1990): Ecology, abundance and distribution of fish life in the north western Himalayan streams. In: Recent trends in limnology: V.P. Agarwal, P. Das (Eds.). Society of Biosciences, Muzaffarnagar, India. PP 287-299.

Sehgal, K.L. (1999): Coldwater fish and fisheries in the Indian Himalayas: Rivers and streams. FAO fisheries Technical paper, Rome. Pp. 41-63.

Shapiro, J. (1973): Blue-green algae: why they become dominant. *Science.*, 179: 382-384.

Sharma, N. (1978): A note on fresh water cladocerans from West Bengal, Bangladesh. J. Zool., 6: 149-151.

Sharma, R.C. (1985): Seasonal abundance of phytoplankton in the Bhairathi River, Garhwal Himalaya. Indian J. Ecol., 12(1): 157-160.

Sharma, A., Sharma, R.C. and Anthwal, A. (2007): Monitoring Phytoplanktonic diversity in the hill stream Chandrbhaga of Garhwal Himalaya. *Life Science J.*, 4(1): 80-84.

Shelford, V.E. (1911): Ecological succession: stream fishes and the methods of physiographic analysis. *Biol. Bull.*, 21: 9-35.

Shulman, M.J. (1984): Resource limitation and recruitment patterns in a coral reef fish assemblage. *J. Exp. Mar. Biol. Ecol.,* 74: 85-109.

Silas, E.G. (1951): On a collection of fishes from Anamalai and Nelliampathi hill ranges (Western Ghats) with notes on its zoogeographical significances. *J. Bombay Nat. Hist. Soc.,* 49: 670-681.

Singh, R.K. and Desai, V.R. (1980): Limnological observations on Rihand reservoir. III primary Productivity. J. Inland Soc. India, Vol. 12, No.2: 63-68.

Singh, B.K., Singh, S.K. and Singh, B.B. (2002). Distribution and seasonal abundance of benthic macro invertebrates in the Bisar sarovar of Gaya, Bihar. *Him.J.Env.Zool.,* Vol. 16(2), pp. 139-150.

Singh, H.P, Choudhry, M. and Kalekaur.V. (1982): Seasonal and diurnal changes in physico-chemical features of the river Brahmaputra at Gauhati, *Ind.J.Zool.,* 2(82):77-84.

Singh, D.K. and Singh A.K. (1985): Zooplankton density in relation of physico-chemical factors during diet cycle river Ganga at Bhagalpur, Bihar. Environ, and Ecol., 3(1): 231-234.

Singh, S.S. and Singh, L.A. (2000): Studies on the physico-chemical characteristics (12 hours variation) of a major water body of Imphal District. J. Environ. Biol., 21 (3): 273-274.

Singh, H.R., Nautiyal, P., Dobriyal, A.K., Pokhriyal R.C., Negi, M., Baduni, V., Nautiyal, R. and Agarwal, N.K. (1994): Water quality of river Ganga (Garhwal Himalaya). *Acta. Hydrobiologia.,* 36(1): 3-15.

Sivakumar, A.A., Arunadevi, P and Aruchami, M. (2003): Studied the water quality of the river Ambarampalayam, Coimbatore District, Tamilnadu. *Nature Environment and Pollution Technology,* Vol.2, No.3, pp. 305-308.

Sladecek, V. (1973): System of water quality from the biological point of view. *Arch. Hydrobiol. Ergebn. Limnol.,* 7: 1-218.

Smock, L.A. (1994): Movements of invertebrates between stream channels and forested flood plains. *J. N. Am. Benthol. Soc.,* 13(4): 524-531.

Sobezak, W.V. and Burton, T.M. (1996): Epilithic bacterial and algal colonization in a stream run, riffle and pool: a test of biomass covariation. *Hydrobiologia.,* 332: 159-166.

Soininen, J. and Kononen, K. (2004): Comparative study of monitoring South-Finnish Rivers and streams using macroinvertebrates and benthic diatoms community structure. *Aquatic Ecology.,* 38: 63-75.

Sousam W.P. (1984): The role of disturbance in natural communities. *Ann. Review Ecol. Syst.,* 15: 353-391.

Statzner, B., Gore, J.A. and Resh, V.H. (1988): Hydraulic stream ecology: observed patterns and potential application. *J.N. Am. Benthol. Soc.,* 7: 307-360.

Stevenson, R.J. (1996): Algal ecology in freshwater benthic habitats. In: R.G. Stevenson *et al.* (eds.), Algal Ecology. Academic Press. San Diego, CA: 3-30.

Stein, D.L., Tissot, B.N., Hixon, M.A. and Barss, W. (1992): Fish-habitat associations on deep reef at the edge of the Orego eontinental shelf. U.S. *Fish. Bull.,* 90: 540-551.

Steinman, A.D., McIntire, D.C., Gregory, S.V. and Lamberti, G.A. (1989): Effects of irradiance and grazing on lotic algal assemblages. *J. Phycol.,* 25: 478-485.

Sullivan, K. (1986): Hydraulics and fish habitat in relation to channel morphology. Docotral Dissertation, Baltimore. Johns Hopkins University.

Sukumaran, P.K. (2002): Coliform bacteria an index of aquatic pollution. *Trends in life Sciences* (India), 17(2), 19-24.

Szedlmayer, S.T. and Able, K.W. (1996): Patterns of seasonal availability and habitat use by fishes and decapod crustaceans in a southern New Jersey estuary. *Estuaries.,* 19: 697-709.

Thomas, G.W. and Crutchfield, J.D. (1974): Nitrate-nitrogen and phosphorus contents of streams draining small agricultural catchments in Kentucky. *J. Environmental Quality.* 3: 46-49.

Traaen, T.S. and Lindstrom, E.A. (1983): Influence of current velocity on periphyton distribution. In: R.G. Wetzel(Ed.). Periphyton of freshwater ecosystem. Development of hydrobiologia, Dr. W. Junk Publishers. The Hague: 121-134.

Tuchman, N.C. and King, R.H. (1993): Changes in mechanism of summer detritus processing between wooden agricultural sites in a Michigan headwater stream. *Hydrobiologia.,* 218: 115-127.

Unni, K.S. (1982): Limnological studies on Sumpna reservoir, Betul, M.P., Proc. Nat. Sci Acad., India 52(B), IV.

Vadas, Jr. R.L. (1994): Habitat tools, for assessing in stream-flow needs for fishes in the upper Roanoke River, Virginia. Doctoral dissertation, Blacksburg: Virginia Polytechnic Institute and State University.

Vadas, Jr. R.L. and Orth, D.J, (1998): Use of physical variables to discriminate visually determined mesohabitat types in North American Streams. *Rivers.,* 6(3): 143-159.

Vavilova, V.V. and Lewis, W.M. (1999): Temporal and altitudinal variations in the attached algae of mountain streams in Colorado, *Hydrobiologia.,* 390: 99-106.

Velz, C.J. (1947): Factory influencing self purification and their relation to pollution abatement (I) S*ewage work J.,* 19:623-644.

Vinogrador, M.E., Shushkine, E.A., Nezlin, N.P. and Genrikh, N.A. (1998): Vertical distribution of zooplankton in the frontal zone of the Gulf stream and Labrador current. *J. Plankt. Res.,* 20(1): 85-103.

Ward, J.V. (1989): The four dimensional nature of lotic ecosystem. *J.N. Am. Benthol. Soc.*, 8: 2-8.

Watanbe, T., Asai, K. and Houki, A. (1986): Numerical estimation to organis pollution of flowing water by using epilithic diatom assemblage-Diatom Assemblage Index (DAIpo) Sci. *Total Environ.,* 55: 209-218.

Webster, M.M. and Hart, P.J.B. (2004): Substrate discrimination and preference in foraging fish. *Animal Behaviour.,* 68: 1071-1077.

Welch, P.S. (1948): Limnological Methods, the Blakiston. Co. Philadelphia, 1-381.

Welch, E.B., Jacoby, J.M., Horner, R.R. and Seelay, M.R. (1988): Nuisance biomass levels of periphytic algae in streams. *Hydrobiologia.,* 157: 161-168.

Wentworth, C.K. (1922): A scale of grade and class for elastic sediments. *Journal of Geology.,* 30: 377-392.

Wilhm, J.L. and Dorris, T.C. (1968): Biological parameters for quality criteria. *Bioscience.,* 18: 477-481.

Wasson, J.G., Bonnard, R. and Maridet, L. (1995): Responses globales des invertebres benthiques aux conditions d'habitat physique dars des cours d'eas salmonicales: poisons. *B. Fr. Peche, Piscie.,* 337/338/339: 291-299.

Williams, D. McB. and Sale, P.F. (1981): Spatial and temporal patterns of recruitment of juvenile coral reef fishes to coral habitats within 'One Tree Lagoon Great Barrier Reef'. *Mar. Biol.,* 65: 245-253.

Wootton, R.J. and Gee, J.H.R. (2001): Conceptual Plant Systematic, 2nd edition. Andrews University Press, Berrien Springs, MI. 619 pp.

Wootton, J.T. and Power, M.E. (1993): Productivity, consumers and the structure of a river food chain. Proceedings of the National Academy of Science, USA., 90: 1384-1387.

Wren, R.C. (1956): Potter's new cyclopaedia of botanical drugs and preparations. Enlarged by R.W. Wren. Sir Isaac Pitmen and Sons Ltd. London.

Wright, J.P. and Flecker, A.S. (2004): Deforesting the riverscape: The effects of wood on fish diversity in a Venezuelan piedmont stream. *Biological Conservation.,* 120:443-451.

Young, T.C. and Kign, D.L. (1973): A rapid method of wuantifying algal carbon uptake kinetics. *Limnol. Oceanogr.,* 18: 978-981.

Zafar, A.R. (1966): Limnology of the Hussain Sagar Lake, Hyderabad, India, *Phykos,* 5(1&2): 115-126.

Zafar, A. and Sultana, N. (2005): Zooplankton and micro-invertebrates of River Ganga at Kanpur. *Uttar Pradesh J. Zool.,* 25(1): 63-66.

INDEX